Alexander Lygin

Novel Syntheses of Nitrogen Heterocycles from Isocyanides

Alexander Lygin

Novel Syntheses of Nitrogen Heterocycles from Isocyanides

Südwestdeutscher Verlag für Hochschulschriften

Impressum/Imprint (nur für Deutschland/ only for Germany)
Bibliografische Information der Deutschen Nationalbibliothek: Die Deutsche Nationalbibliothek verzeichnet diese Publikation in der Deutschen Nationalbibliografie; detaillierte bibliografische Daten sind im Internet über http://dnb.d-nb.de abrufbar.

Alle in diesem Buch genannten Marken und Produktnamen unterliegen warenzeichen-, marken- oder patentrechtlichem Schutz bzw. sind Warenzeichen oder eingetragene Warenzeichen der jeweiligen Inhaber. Die Wiedergabe von Marken, Produktnamen, Gebrauchsnamen, Handelsnamen, Warenbezeichnungen u.s.w. in diesem Werk berechtigt auch ohne besondere Kennzeichnung nicht zu der Annahme, dass solche Namen im Sinne der Warenzeichen- und Markenschutzgesetzgebung als frei zu betrachten wären und daher von jedermann benutzt werden dürften.

Verlag: Südwestdeutscher Verlag für Hochschulschriften Aktiengesellschaft & Co. KG
Dudweiler Landstr. 99, 66123 Saarbrücken, Deutschland
Telefon +49 681 37 20 271-1, Telefax +49 681 37 20 271-0
Email: info@svh-verlag.de
Zugl.: Göttingen: Georg-August-Universität, Diss., 2009

Herstellung in Deutschland:
Schaltungsdienst Lange o.H.G., Berlin
Books on Demand GmbH, Norderstedt
Reha GmbH, Saarbrücken
Amazon Distribution GmbH, Leipzig
ISBN: 978-3-8381-1754-6

Imprint (only for USA, GB)
Bibliographic information published by the Deutsche Nationalbibliothek: The Deutsche Nationalbibliothek lists this publication in the Deutsche Nationalbibliografie; detailed bibliographic data are available in the Internet at http://dnb.d-nb.de.

Any brand names and product names mentioned in this book are subject to trademark, brand or patent protection and are trademarks or registered trademarks of their respective holders. The use of brand names, product names, common names, trade names, product descriptions etc. even without a particular marking in this works is in no way to be construed to mean that such names may be regarded as unrestricted in respect of trademark and brand protection legislation and could thus be used by anyone.

Publisher: Südwestdeutscher Verlag für Hochschulschriften Aktiengesellschaft & Co. KG
Dudweiler Landstr. 99, 66123 Saarbrücken, Germany
Phone +49 681 37 20 271-1, Fax +49 681 37 20 271-0
Email: info@svh-verlag.de

Printed in the U.S.A.
Printed in the U.K. by (see last page)
ISBN: 978-3-8381-1754-6

Copyright © 2010 by the author and Südwestdeutscher Verlag für Hochschulschriften Aktiengesellschaft & Co. KG and licensors
All rights reserved. Saarbrücken 2010

Novel Syntheses of Nitrogen Heterocycles from Isocyanides

DISSERTATION

zur Erlangung des
mathematisch-naturwissenschaftlichen Doktorgrades
"Doctor rerum naturalium"
der Georg-August-Universität Göttingen

vorgelegt von

Alexander Lygin

aus

Krasnokamensk, Russland

Göttingen 2009

D7

Referent: Prof. Dr. A. de Meijere

Korreferent: Prof. Dr. U. Diederichsen

Tag der mündlichen Prüfung:

Die vorliegende Arbeit wurde in der Zeit von November 2006 bis Oktober 2009 im Institut für Organische und Biomolekulare Chemie der Georg-August-Universität Göttingen durchgeführt.

Für die Überlassung des Themas, die hilfreichen Diskussionen und Anregungen sowie die ständige Unterstützung während der Arbeit möchte ich meinem Lehrer, Herrn Prof. Dr. A. de Meijere, ganz herzlich danken.

Der Degussa(Evonik)-Stiftung danke ich für die Gewährung eines Promotionsstipendiums.

Dedicated to Tonja and Masha

Table of Contents

A. INTRODUCTION AND BACKGROUND ... 5

1. Isocyanides in Organic Synthesis ... 5
2. Cyclizations of Metallated Isocyanides ... 8
   2.1. α-Metallated Methyl Isocyanides ... 8
   2.2. α-Metallated *ortho*-Methylphenyl Isocyanides ... 19
   2.3. Other Metallated Isocyanides ... 25
3. Addition to the Isocyano Group Followed by a Cyclization ... 26
   3.1. Non-Catalyzed Processes ... 26
   3.2. Transition Metal-Catalyzed Processes ... 32
4. Goals of this Study ... 35

B. MAIN PART ... 36

1. Oligosubstituted Pyrroles Directly from Substituted Methyl Isocyanides and Acetylenes ... 36
   Background and Preliminary Considerations ... 36
   Synthesis of 2,3,4-Trisubstituted and 2,4-Disubstituted Pyrroles ... 36
   Kinetic Studies ... 43
   Synthesis of 2,3-Disubstituted Pyrroles ... 44
   Mechanistic Considerations ... 52
   Conclusion ... 54
2. *ortho*-Lithiophenyl Isocyanide: A Versatile Precursor to 3*H*-Quinazolin-4-ones and 3*H*-Quinazolin-4-thiones ... 55
   Background and Preliminary Considerations ... 55
   Synthesis of 2-Substituted Phenyl Isocyanides by Reaction of *ortho*-Lithiophenyl Isocyanide with Electrophiles ... 56
   Synthesis of Substituted 3*H*-Quinazolin-4-ones and 3*H*-Quinazolin-4-thiones ... 58
   Conclusion ... 61
3. Reactions of *ortho*-Lithiophenyl (-Hetaryl) Isocyanides with Carbonyl Compounds – Rearrangements of 2-Metallated 4*H*-3,1-Benzoxazines ... 62
   Background and Preliminary Considerations ... 62

Reactions of *ortho*-Lithiophenyl (-Hetaryl) Isocyanides with Carbonyl Compounds 62

Copper(I)-catalyzed Cyclizations of Isocyanobenzyl alcohols **204**. 67

Novel Rearrangements of 2-Metallated 4*H*-3,1-Benzoxazines 70

Mechanistic Considerations 72

Conclusion 74

4. Synthesis of 1-Substituted Benzimidazoles from *o*-Bromophenyl Isocyanide and Amines 75

Background and Preliminary Considerations 75

Optimization of the Reaction Conditions for the Synthesis of 1-Benzylbenzimidazole 76

Scope and Limitations of the Synthesis 78

Conclusion 82

C. EXPERIMENTAL SECTION 83

General 83

Experimental Procedures for the Compounds Described in Chapter 1 "Oligosubstituted Pyrroles Directly from Substituted Methyl Isocyanides and Acetylenes" 85

Experimental Procedures for the Compounds Described in Chapter 2 "*ortho*-Lithiophenyl Isocyanide: A Versatile Precursor for 3*H*-Quinazolin-4-ones and 3*H*-Quinazolin-4-thiones" ... 99

Experimental Procedures for the Compounds Described in Chapter 3 "Reactions of *ortho*-Lithiophenyl (-Hetaryl) Isocyanides with Carbonyl Compounds – Rearrangements of 2-Metallated 4*H*-3,1-Benzoxazines" 111

Experimental Procedures for the Compounds Described in Chapter 4 "Synthesis of 1-Substituted Benzimidazoles from *o*-Bromophenyl Isocyanide and Amines" 130

D. SUMMARY AND OUTLOOK 139

E. REFERENCES AND COMMENTS 145

F. REPRESENTATIVE ^{1}H AND ^{13}C SPECTRA OF THE PREPARED COMPOUNDS 161

List of Abbreviations

Ac	=	Acetyl
acac	=	Acetylacetonato
AIBN	=	Azabisisobutyronitrile
BINAP	=	2,2'-Bis(diphenylphosphino)-1,1'-binaphthyl
Bn	=	Benzyl
Bu	=	Butyl
CHIRAPHOS	=	(2R,3R)-(+)-Bis(diphenylphosphino)butane
Cp	=	Cyclopentadienyl
nCR	=	n-Component Reaction
DIOP	=	O-Isopropyliden-2,3-dihydroxy-1,4-bis(diphenylphosphino)butan
DBU	=	1,8-Diazabicyclo[5.4.0]undec-7-ene
DCM	=	Dichloromethane
DDQ	=	2,3-Dichloro-5,6-dicyanobenzoquinone
DMF	=	N,N-Dimethylformamide
DMSO	=	Dimethyl sulfoxide
dppp	=	1,3-Bis(diphenylphosphino)propane
E^+ or El	=	Electrophile
Et	=	Ethyl
ee	=	Enantiomeric excess
EWG	=	Electron-withdrawing group
cHex	=	Cyclohexyl
HMPA	=	Hexamethylphosphortriamide
KHMDS	=	Potassium bis(trimethylsilyl)amide [$KN(SiMe_3)_2$]
LA	=	Lewis acid
LDA	=	Lithium diisopropylamide
LiTMP	=	Lithium 1,1,6,6-tetramethylpiperidide
Me	=	Methyl
MTBE	=	Methyl *tert*-butyl ether
NP	=	Nanoparticles
Nu$^-$ or NuH	=	Nucleophile
Ph	=	Phenyl
1,10-Phen	=	1,10-Phenanthroline
Pr	=	nPropyl

cPr	=	Cyclopropyl
THF	=	Tetrahydrofuran
TBDMS	=	*tert*-Butyldimethylsilyl
TMEDA	=	N,N,N',N'-Tetramethylethylenediamine
pTol	=	pTolyl
TosMIC	=	pToluenesulfonylmethyl isocyanide

A. Introduction and Background

1. Isocyanides in Organic Synthesis

Isocyanides have been first described independently by Gautier[1] to be formed in the reaction of silver cyanide with alkyl iodides and by Hofmann[2] upon treatment of aniline with chloroform in the presence of potassium hydroxide (the so-called carbylamine reaction). Because of the extremely unpleasant odor of the simplest (and the most volatile) isocyanides, efficient methods for their synthesis have not been developed for a long time, and therefore these compounds have long been underinvestigated. The chemistry of isocyanides received a significant boost when reliable methods for the synthesis of isocyanides, on a wide scope, e.g. the dehydration of formamides[3] and the carbylamine reaction of amines employing phase-transfer catalysis[4] appeared in the literature. The carbon atom of the isocyano group often exhibits carbene-like reactivity that is reflected in the resonance structure **1a** (Scheme 1). Conversely, the linear structure of isocyanides is well represented by the dipolar resonance structure **1b**. Such unique properties of the isocyano group, which may function both as an electrophile and as a nucleophile coupled with the now easy availability of a wide range of isocyanides have turned these compounds into indispensable building blocks for organic synthesis.[5]

$$R-N=C: \longleftrightarrow R-\overset{+}{N}\equiv C^-$$
$$\text{1a} \qquad\qquad \text{1b}$$

Scheme 1.   Resonance structures of isocyanides.

The diversity of transformations, which isocyanides can undergo, includes various isocyanide-based multicomponent reactions, e.g. the Ugi and Passerini reactions (Scheme 2),[6] other (Lewis acid-catalyzed) cocyclizations utilizing isocyanides as one-carbon donor (e.g. depicted on Scheme 3)[7] as well as their transition-metal catalyzed insertions,[8,9] oligo- and polymerizations.[10] Arguably the most important applications of isocyanides are toward the synthesis of various heterocycles.
Isocyanides are also well-known to participate in different types of radical processes to provide various heterocycles. Once generated, the radical intermediates readily undergo addition to an isocyano group to produce the corresponding imidoyl radicals, which in some cases are capable of subsequent cyclizations to give heterocyclic compounds.

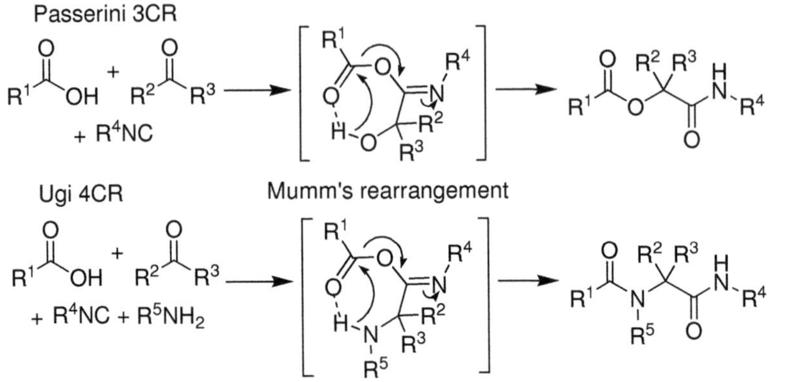

Scheme 2. The three-component Passerini and the four-component Ugi reaction.

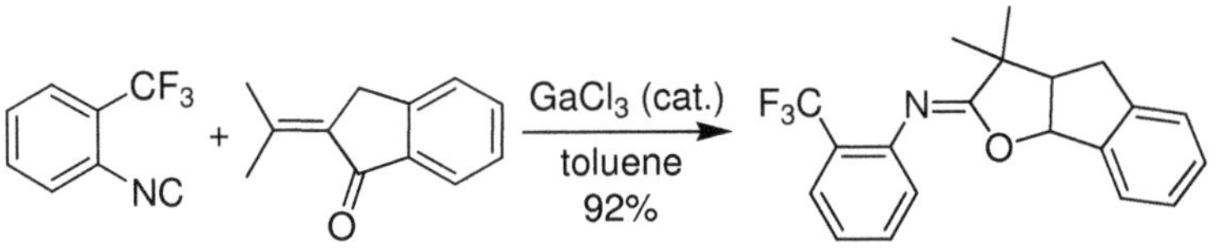

Scheme 3. An example of a formal [4+1]-cycloaddition of an α,β-unsaturated carbonyl compound with an isocyanide.[7f]

One of the best known and important processes of this type, which has been developed by Fukuyama et al., is the synthesis of indoles **3** by treatment of *o*-isocyanostyrenes **2** with tri-*n*-butyltin hydride and the radical initiator azobisisobutyronitrile (AIBN) (Scheme 4).[11] The resulting 2-tributylstannyl indoles **3** can be converted into 3-substituted indoles of type **4** simply by acidic workup, but more importantly, they provide a convenient access to various 2,3-disubstituted indoles of type **6** by Stille cross-coupling reactions. The tributylstannyl derivate **3** also reacts smoothly with iodine to provide the 2-iodoindole **5**, another useful substrate for subsequent modifications, , which has been shown to undergo various cross-coupling reactions.[11]

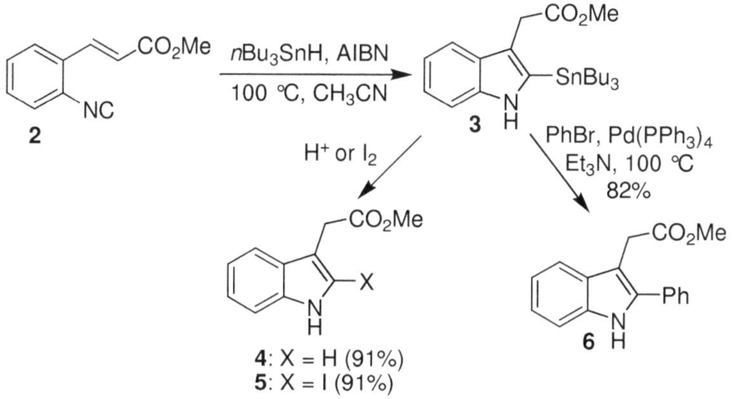

Scheme 4. The Fukuyama's indole synthesis.[11]

Diverse sequential radical cocyclizations with isocyanides, a representative example of which concerns the synthesis of (20S)-camptothecin **8**[12] as depicted in Scheme 5, have previously been reviewed by Curran et al.[13]

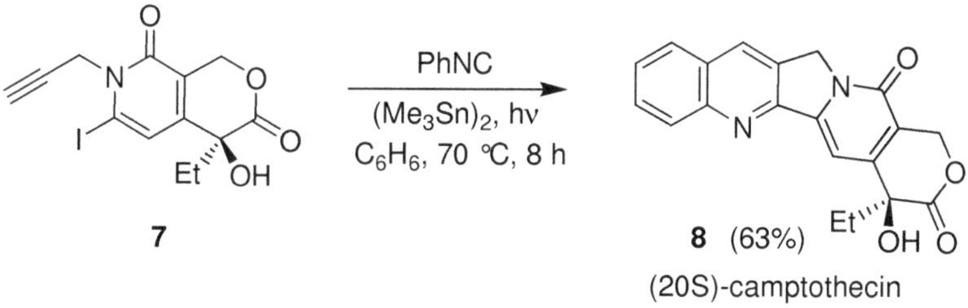

Scheme 5. An example of a sequential radical cocyclization of **7** with phenyl isocyanide. Synthesis of (20S)-camptothecine (**8**).[12]

Two other (non-radical) general types of cocyclizations leading to the formation of heterocycles from isocyanides, should be considered more closely as they are more relevant to the experimental work of this doctoral study, namely: 1) cocyclizations of metallated isocyanides and 2) formal α-additions onto the isocyano group followed by a cyclization. This concise overview might help us to understand that has been previously done in this area and help to imagine new possible directions of development.

2. Cyclizations of Metallated Isocyanides

2.1. α-Metallated Methyl Isocyanides

The electron-withdrawing effect of the isocyano group enhances the acidity of α-C, H bonds, and this was first exploited by Schöllkopf and Gerhart[14] in 1968. Since then, α-metallated methyl isocyanides of type **13** (mostly deprotonated isocyanoacetates) have been shown to participate in various types of cocyclizations leading to different nitrogen- containing heterocycles. Several reviews on this topic had appeared by 1985.[15]

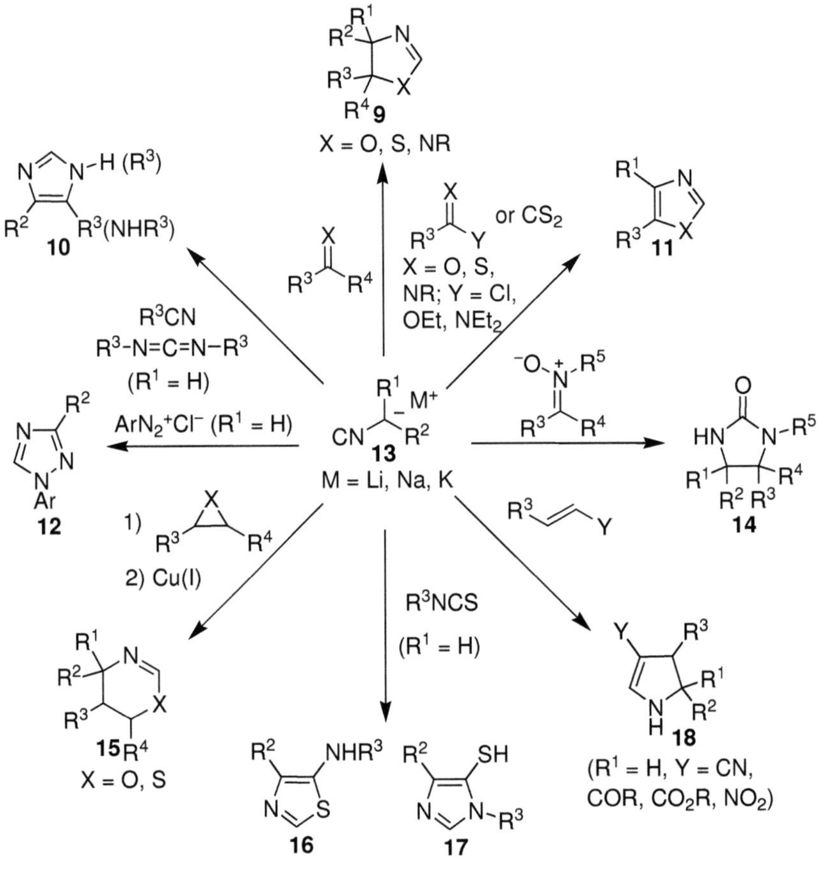

Scheme 6. Various applications of α-metallated substituted methyl isocyanides **13** reviewed previously.[15]

The main types of transformations reported therein as depicted in Scheme 6 include syntheses of 1,3-azoles **10**, **11**, **16**, **17** (azolines **9**), pyrrolines **18**, 1,2,4-tetrazoles **12**, 2-imidazolinones **14**, and 5,6-dihydro-4H-1,3-oxazines (-thiazines) **15**.[15]

One of the most important applications of α-metallated methyl isocyanides is undoubtedly in the preparation of 1,2-disubstituted pyrroles by their reaction with nitroalkenes.[16] In this so-called Barton-Zard pyrrole synthesis the nitro group on the alkene **19** serves two purposes, namely to activate the double bond in **19** toward Michael addition of the deprotonated isocyanide and to provide a leaving group for the conversion of the initially formed 2-pyrroline **21** into a 1H-pyrrole **23** by overall elimination of nitrous acid and subsequent 1,5-sigmatropic hydrogen-shift in the 3H-pyrrole **22** (Scheme 7).

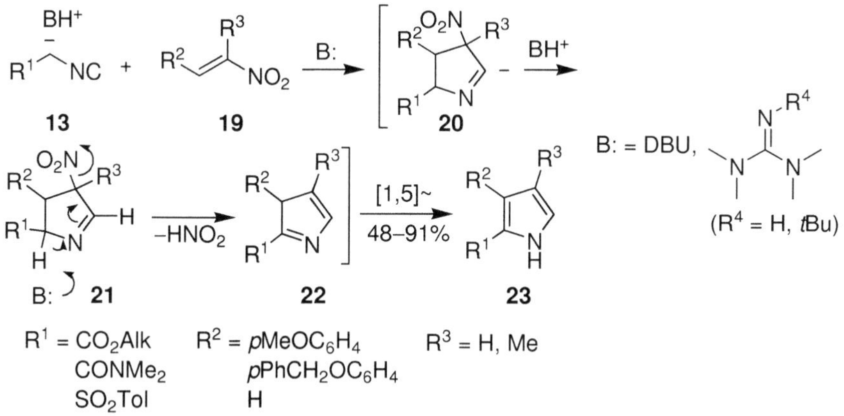

Scheme 7. The Barton-Zard pyrrole synthesis.[16]

The nitroalkanes required for this synthesis are easily accessible by an aldol-type condensation of nitroalkanes with aldehydes (Henry reaction); they can also be generated in situ from O-acetyl-β-hydroxynitroalkanes (Scheme 8, eq. (1)).[16,17] When a non-ionic superbase like **31**, which is about 10^{17} times more basic than 1,8-Diazabicyclo[5.4.0]undec-7-ene (DBU) is employed instead of DBU, the respective pyrroles are obtained in excellent yields (Scheme 8, eq (2)).[18] The same base **31**, has been shown also to be superior over DBU in the synthesis of oxazoles **30** by reaction of acid chlorides **29** and anhydrides with methyl isocyanoacetate (**25-Me**) providing the products fast and in almost quantitative yields.[18]

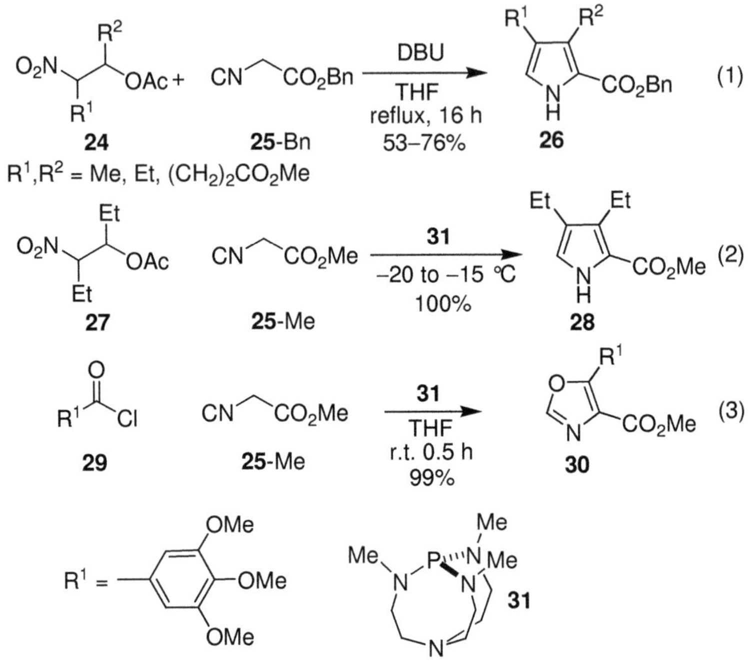

Scheme 8. In situ generation of nitroalkenes in the Barton-Zard pyrrole synthesis. Some applications of the superbase **31**.[16,18]

The quality and the type of the solvent, particularly the absence of radical inhibitors such as BHT which is routinely added to commercial THF, have been shown to influence the rate of the reaction as well as the pyrrole yields.[19] *tert*-Butyl methyl ether (MTBE) has been found to be better than THF in this reaction.

The reaction of ethyl isocyanoacetate **25-Et** with certain nitroaromatic compounds, e. g. 9-nitrophenanthrene (**32**), also provided the corresponding pyrrole **33** fused to a phenanthrene moiety (Scheme 9).[20] Polycyclic aromatic nitro compounds with decreased aromaticity gave the corresponding arene-annelated pyrroles in good yields while simple nitroarenes such as nitronaphtalene and nitrobenzene turned out to be less efficient or even failed in this reaction.[20]

Alternatively to nitroalkenes, α,β-unsaturated phenylsulfones **35** can be employed in the synthesis of pyrroles **36** with the same substitution pattern as in the Barton-Zard method (Scheme 10).[21] This reaction proceeds with elimination of phenylsulfinic acid PhSO$_2$H.

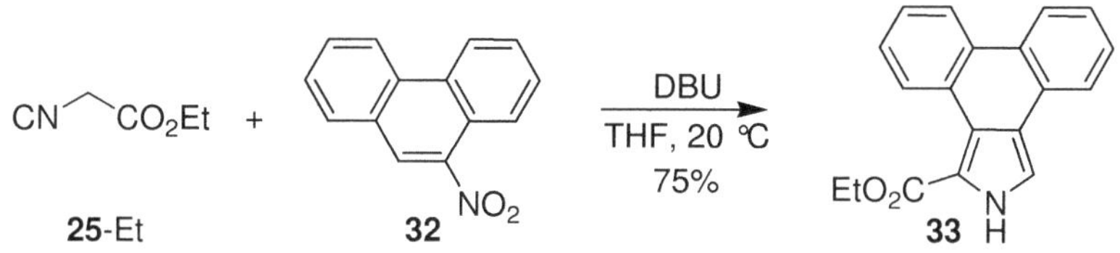

Scheme 9. Synthesis of pyrrole **33**.[20]

The phenylsulfones of type **35** are easily accessible e. g. by sulfenohalogenation of alkenes with subsequent β-elimination of hydrogen halide from the resulting adducts. α,β-Unsaturated nitriles, which conveniently prepared by condensation of substituted aryl- acetonitriles with aldehydes, in turn have been shown to react with deprotonated isocyanoacetates **25** to provide, after elimination of cyanide, 3,4-diarylpyrrole-2-carboxylates in moderate yields.[22]

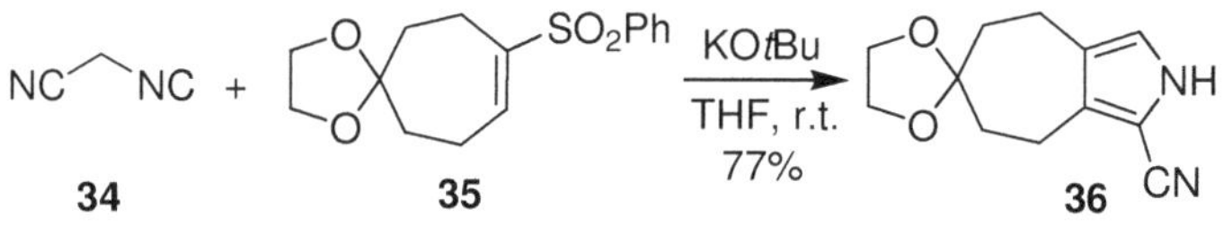

Scheme 10. Synthesis of pyrrole **36** from α,β-unsaturated sulfone **35** and isocyanide **34**.[21]

Polarized ketene S,S-dithioacetals of type **37** or N,S-acetals **38** (Fig. 1) represent further suitable counterparts for activated methyl isocyanides in the synthesis of 2,3,4-trisubstituted pyrroles.[23] These base-induced reactions proceed with elimination of methylthiolate and loss of the respective substituents R^1.[23]

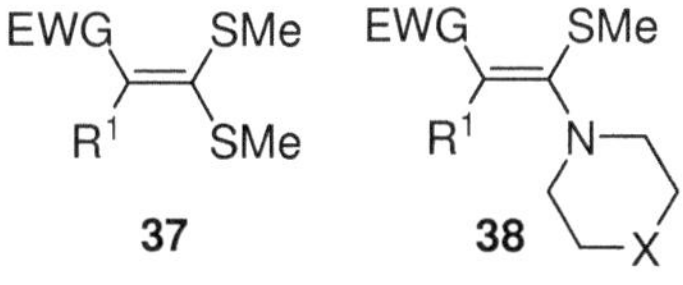

EWG = NO$_2$, CN, COMe, COPh, CO$_2$Et
R^1 = H, CO$_2$Et, COMe, COPh
X = O, NCO$_2$Et, NCH$_2$Ph

Figure 1. Polarized ketene S,S-dithioacetals **37** and N,S-acetals **38**.[23]

The Barton-Zard methodology has been employed in various natural product syntheses, such as that of pyrrolostatin and its analogues[24] as well as chromophores for biological systems.[25] Importantly, the pyrroles synthesized from α,β-unsaturated nitroalkenes or phenylsulfones posses a substitution pattern perfect for the construction of porphyrines.[20, 21d,e,f, 26] Thus, reduction of the ester group at position 2 of the pyrrole **39**, succeeding acid-catalyzed cyclizing condensation with an excess of methylal (formaldehyde dimethylacetal) and subsequent oxidation led to octaethylporphyrin **40** in 69% yield over three steps (Scheme 11).[18, 26]

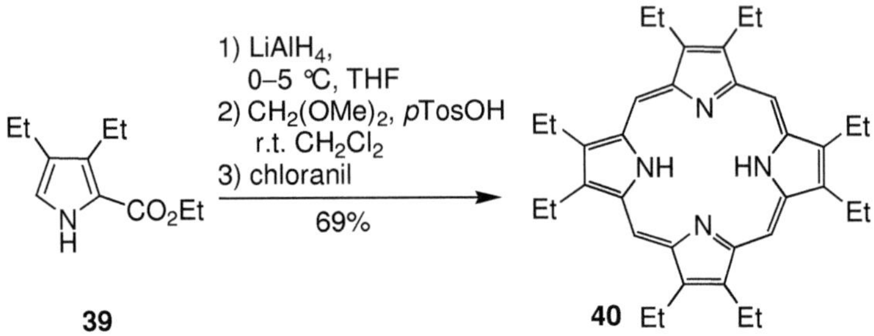

Scheme 11. Synthesis of octaethylporphyrin **40** from the pyrrole **39**.[18]

The most frequently used α-isocyanoalkanoic acid derivatives contain ester groups as acceptors and are easily accessible from the corresponding amino acids. Some acceptor substituents on methyl isocyanides, e. g. the tosyl group, capable of further elimination under basic conditions, may bring some synthetic advantages toward particular heterocycles from isocyanides. Tosylmethyl isocyanide (TosMIC, **41-H**)[27] introduced in organic synthesis and employed for various purposes by van Leusen, has become a classical reagent for the construction of 1,3-azoles and pyrroles.[28] Thus, it reacts under basic conditions (with elimination of TosH): with aldehydes to provide oxazoles;[29] with aldimines to give imidazoles;[30,31] with acceptor-substituted alkenes to give pyrroles (Scheme 12).[32] The latter reaction, known as the van Leusen pyrrole synthesis, is of particular importance, as pyrroles are widespread among naturally occurring biologically active substances and their synthetic analogues. Pyrroles thus prepared from isocyanides **41-R**, can be further elaborated. Thus, α-trimethylstannyl-substituted TosMIC (**41-SnMe₃**) employed in this reaction, provides an access to 2-(trimethylstannyl)pyrroles, which could be further derivatized e. g. by Stille cross-coupling reactions with aryl bromides.[33]

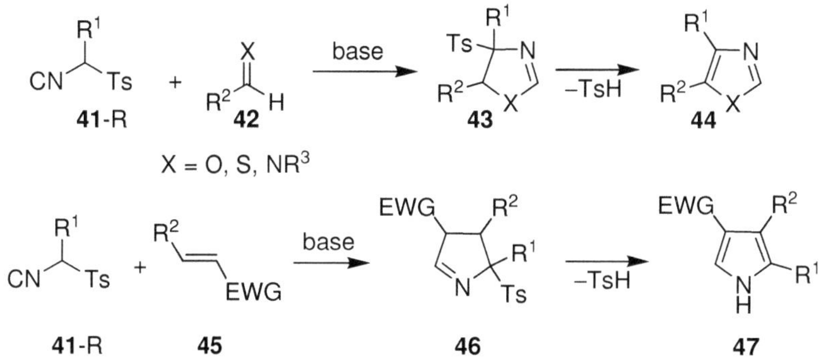

Scheme 12. Synthesis of various 1,3-azoles from tosylmethyl isocyanide and its derivatives **41**-R (R = H, TosMIC).[28–32]

Interestingly, mono- and 1,2-disubstituted arylalkenes (preferably with electron-withdrawing substituents) have been shown also to provide 3-aryl- or 3,4-diarylsubstituted pyrroles, respectively, in moderate to good yields by the reaction of TosMIC in the presence of NaOtBu as a base in DMSO.[34]

A base-induced reaction of 1-isocyano-1-tosyl-1-alkylidene methyl isocyanides **51** with unsaturated compounds of type **49** furnished azoles **50** capable to undergo a subsequent pericyclic reaction and aromatization by means of DDQ to give various benzoannelated heterocycles: indoles **52**, benzimidazoles **54** and benzoxazoles **55**, respectively (Scheme 13).[35] Apparently, a strong base such as potassium *tert*-butoxide deprotonates the isocyanide **51** to furnish the isocyanoallyl anion **48**, which cocyclizes with acceptor-substituted alkenes **49** (X = CHCOR^3), aldehydes (X = O) or imines (X = NR) to provide the corresponding azoles.

Another example of an acceptor-substituted methyl isocyanide in which the acceptor is a good leaving group, benzotriazol-1-yl-methyl isocyanide (BetMIC), has been reported by Katritzky et al. to be sometimes superior over TosMIC in the synthesis of oxazoles, imidazoles and pyrroles.[36]

In addition to base-mediated reactions, the catalytic versions of some of the corresponding cocyclizations of substituted methyl isocyanides with unsaturated compounds have been intensively investigated. Copper(I), silver(I) and gold(I) salts are most frequently used catalysts for the aforementioned syntheses of heterocycles. Thus, Cu(I)-, Ag(I)- or Au(I)-catalyzed reactions of substituted methyl isocyanides with aldehydes (ketones),[37]

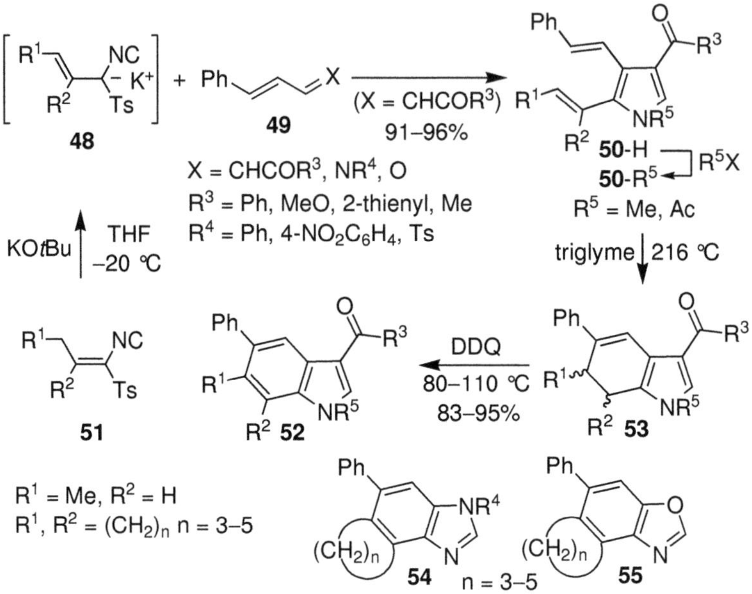

Scheme 13. Synthesis of indoles **52**, benzimidazoles **54** and benzoxazoles **55** by sequential construction of an azole ring and a benzene ring.[35]

imines,[38] as well as various Michael acceptors[39] have been reported. Such catalytic variants have some obvious advantages over conventional (base-mediated) reactions, i. e. atom economy,[40] and the possibility to use base-sensitive substrates as well as to be able to obtain the respective products diastereo- or even enantioselectively. The asymmetric synthesis of synthetically useful 4,5-disubstituted 2-oxazolines **57** by an aldol-type condensation of aldehydes with substituted methyl isocyanides containing an electron-withdrawing group has first been reported by Ito et al. in 1986.[41a] Thus, in the presence of 1 mol% of a Au(I) complex with chiral bis(diphenylphosphino)ferrocene ligands of type **58**, the reaction of methyl isocyanoacetate (**25-Me**) with aldehydes gave the respective *trans*-disubstituted (4*S*, 5*R*)-oxazolines in high yields (83–100%) diastereo- and enantioselectively (Scheme 14).[41] Isocyanomethylcarboxamides,[42 a, d] -phosphonates[42b] and α-ketoesters[41f] have also successfully been employed in this cocyclization while the reaction with other α-substituted methyl isocyanocarboxylates proceeded notably slower than with methyl isocyanoacetate (**25-Me**) and sometimes with decreased stereo- and enantioselectivity.[41c,d] The silver complexes with ligands of type **58** were found to be superior over their gold(I) analogues for the reaction of aldehydes with TosMIC[42c] and provided the corresponding *trans*-(4R, 5R)-5-alkyl-4-tosyl-2-oxazolines in excellent yields with high degrees of diastereo- and enantioselectivity (up to 86% ee).

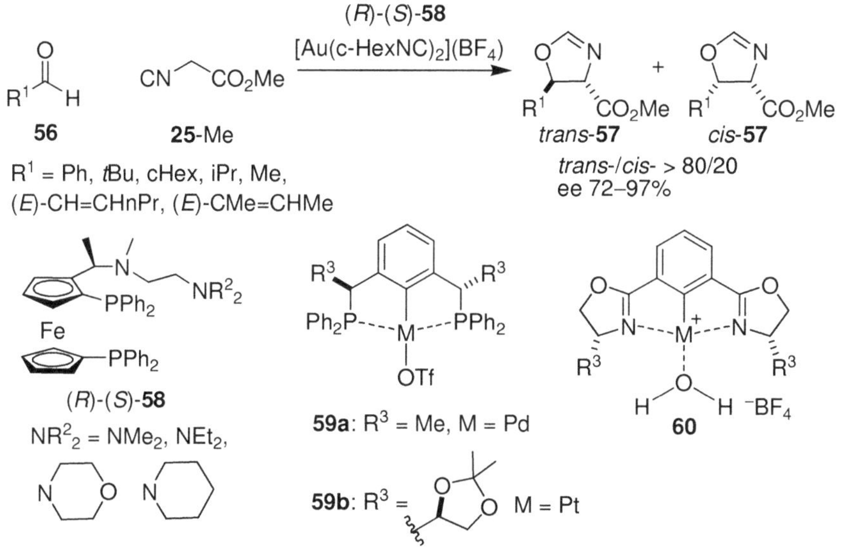

Scheme 14. Asymmetric synthesis of 4,5-disubstituted 2-oxazolines **57**.[41, 44]

The mechanism of this reaction has been extensively studied in order to understand the mode of action of the catalyst and the reason for its high stereoselectivity.[43] It has been shown, that the "internal cooperativity" of both central and planar chirality of the ligand **58** plays a crucial role in the high diastereo- and enantioselectivity of the reaction observed. Thus, other combinations of both chirality types have been shown to be less efficient. The secondary interactions between a pendant amine and substrate are also crucial as metal complexes with other chiral bidentate phosphine ligands, e. g. CHIRAPHOS, DIOP, and BINAP lead to almost racemic oxazolines. A mechanistic explanation for this fact is that enolates derived from isocyanoactetate in this aldol-type reaction are placed too far away from the chiral pocket formed by such ligands, so that they cannot control the stereochemical outcome of the reaction.

Some Pd(II), Pt(II) and Pt(IV) complexes of chiral PCP- and PNP pincer-type ligands with a deeper chiral pocket around the metals have indeed been successfully employed in the asymmetric synthesis of 4,5-disubstituted oxazolines, although with inferior results when compared to the above mentioned Au(I) complexes.[44] Among them, the best diastereo- and enantioselectivities have been observed with depicted in Scheme 14 complexes of type **59a** (*trans*/*cis*: 45/55 to 91/9; *trans*: low ee (<30%); *cis*: 42–77% ee)[44b] **59b** (*trans*/*cis*: 56/44 to 93/7; *cis*: low ee; *trans*: 13–65% ee)[44c] and

60 (reaction with TosMIC: >99% *trans* (4*S*, 5*S*); 25–75% ee; reaction with **25**-Me: low stereoselectivity).[44d]

The Au(I)-catalyzed reaction of alkylisocyanoacetates (**25**-R) with *N*-tosylimines (**61**) afforded the respective *cis*-(4*R*, 5*R*)-2-imidazolines **62** (in contrast to reactions with aldehydes) enantioselectively with the ligand (*R*)-(*S*)-**58** (Scheme 15).[45] Interestingly, the combination of the same ligand (*R*)-(*S*)-**58** with bis-(cyclohexyl isocyanide)gold(I) tetrafluoroborate afforded the respective isomer *trans*-**62** diastereo- and enantioselectively.

cis-2-Imidazolines could also be synthesized diastereoselectively with achiral RuH$_2$(PPh$_3$)$_4$[46] as a catalyst and diastereo-[47] and enantioselectively with some chiral Pd(II)-pincer complexes.[48] *trans*-Stereoselective synthesis of *N*-sulfonyl-2-imidazolines by a Cu(I)-catalyzed reaction of *N*-tosylimines with isocyanoacetates has also been reported.[49]

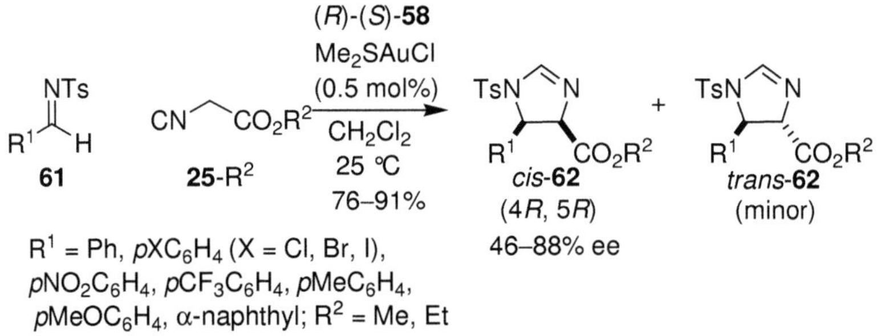

Scheme 15. Asymmetric synthesis of 4,5-disubstituted imidazolines **62**.[45]

Low catalyst loadings and high degrees of diastereo- and enantioselectivity make such aldol-type reactions (especially their Ag(I) and Au(I)/**58**-catalyzed variants discussed above) extremely valuable tools in organic synthesis.

The efficient synthesis of oligosubstituted pyrroles **65** by a formal cycloaddition of isocyanides **63** across the triple bond of electron-deficient alkynes **64** has been reported independently by Yamamoto et al.[50] and by de Meijere et al. (Scheme 16).[51] In our group this reaction has been performed both in the presence of bases such as KO*t*Bu and KHMDS and catalytically (CuSPh, Cu$_2$O and metallic Cu nanoparticles have shown the best results in this case). Importantly, only the base-induced variant allows to efficiently employ substituted methyl isocyanides **63** even without electron-withdrawing groups, e. g. benzyl isocyanide, for the synthesis of pyrroles. Yamamoto et al. have reported similar results on the catalyzed formation of pyrroles **65** with Cu$_2$O/1,10-phenanthroline as the catalytic system of choice. A broad scope of isocyanides **63** and acetylenes **64** have been involved in this catalytic reaction. Recently, a similar solid-phase Cu$_2$O-catalyzed synthesis of 2,3,4-trisubstituted pyrroles **65** by a reaction of polymer-supported acetylenic sulfones with methyl isocyanoacetate (**25-Me**) has been reported.[52]

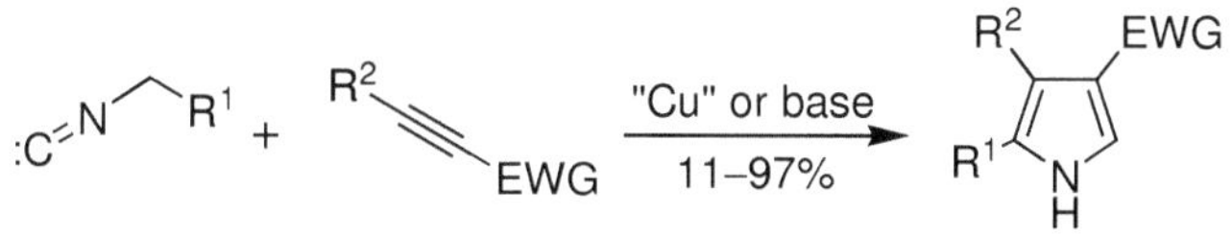

$$:C\!\equiv\!N\text{-}R^1 + R^2\text{-}\!\!\equiv\!\!\text{-}EWG \xrightarrow[11-97\%]{\text{"Cu" or base}} \underset{H}{\underset{|}{R^1\text{-}N}}\text{-pyrrole with } R^2, EWG$$

63
$R^1 = CO_2R^3$ (R^3 = Me, Et, tBu), Ph
$CONEt_2$, CN, $P(O)(OEt)_2$, SO_2Tol

64

65
R^2 = Me, CH_2OMe, cPr, CF_3, Ph, tBu, cHex, N-morpholino, $(CH_2)_4OH$, CO_2Et
EWG = CO_2R^3 (R^3 = Me, Et, tBu), CN, COMe, $CONEt_2$, SO_2Ph, $P(O)(OEt)_2$

Scheme 16. Synthesis of 2,3,4-trisubstituted pyrroles **65** from substituted methyl isocyanides **63** and alkynes **64**.[50a,51]

Yamamoto et al. have also reported the regioselective phosphine-catalyzed formation of pyrroles **66** from the same starting materials **63** and **64** (Scheme 17).[50] This interesting organocatalytic transformation has been found to give best yields in dioxane at 100 °C with bidentate phosphines such as dppp as catalysts. The proposed mechanism includes the addition of a phosphine **68** onto the activated C-C triple bond of an acceptor-substituted alkyne **64** to form a zwitterionic intermediate **70**, which in turn deprotonates the isocyanide **63**, releasing the alkene **71**. The strongly electron-withdrawing phosphonium substituent attached to to the double bond of **71** leads to a reversion of the normal reactivity (*Umpolung*) of this derivative toward a nucleophilic attack of deprotonated methyl isocyanide **13**. Thus, the formal cycloaddition of **13** onto the double bond of **71**, followed by elimination of a phosphine in the first formed intermediate **69** leads to **67** and a [1,5]-hydrogen shift finally provides the pyrroles **66**, the regioisomers of **65**. This method represents an important supplement to the previously discussed synthesis of **65**, although it is applicable only to methyl isocyanides with electron-withdrawing substituents.

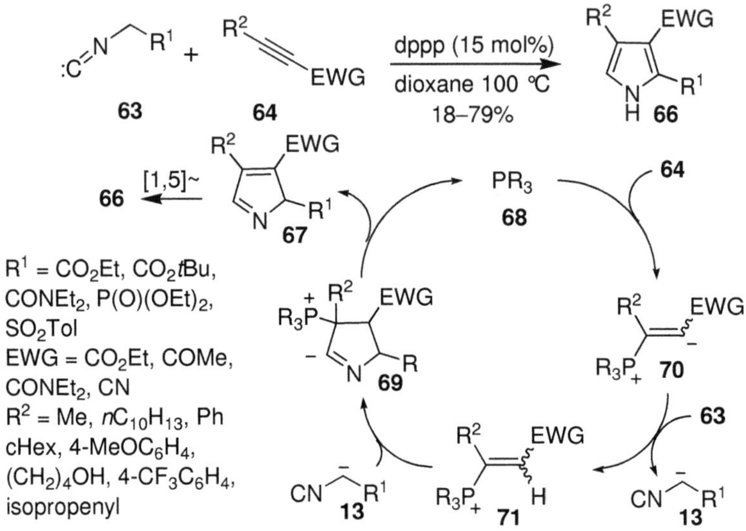

Scheme 17. A plausible mechanism for the phosphine-catalyzed formation of pyrroles **66** from substituted methyl isocyanides **63** and acetylenes **64**.[50]

Substituted methyl isocyanides such as methyl isocyanoacetate (**25-Me**), have been observed to efficiently undergo a dimerization leading to imidazoles under Ag(I), Au(I) or Cu(I) catalysis. [51, 39] The catalytic heterocoupling reaction of two different isocyanides **72-R**^{1} and **34** developed by Yamamoto et al., provided various 1,4-disubstituted imidazoles **73** usually in high yields (Scheme 18).[53] The most efficient catalytic system was found to be Cu_2O/1,10-phenanthroline. Aryl isocyanides **72-R**^{1} with various substituents and some acceptor-substituted methyl isocyanides (**63**) were successfully employed in this transformation, while the reaction of phenyl isocyanide with benzyl isocyanide afforded only traces of the respective imidazoles.

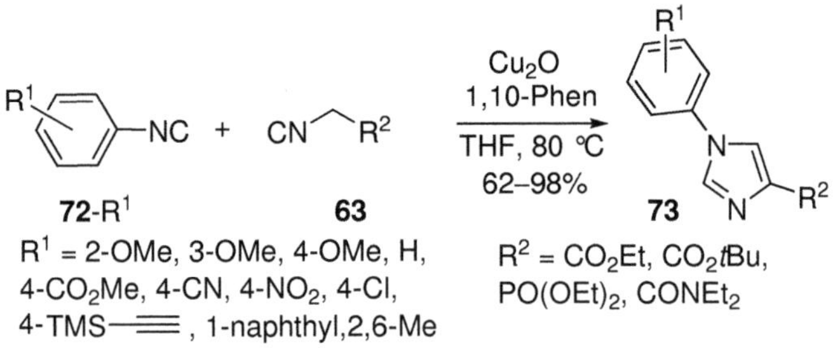

Scheme 18. Cu_2O-Catalyzed synthesis of imidazoles **73** from two different isocyanides **72-R** and **34**.[53]

The rhodiumcarbonyl complex-catalyzed reaction of ethyl isocyanoacetate (**25**-Et) with an excess of a 1,3-dicarbonyl compound **74** (2 equiv.) represents another catalytic approach toward substituted pyrroles (Scheme 19).[54] The reaction of isocyanide **25**-Et with carbonyl compounds produces unsaturated formamides of type **76**, when performed in the presence of a stoichiometric amount of a base such as BuLi or NaH.[55] The same transformation occurs also with $Rh_4(CO)_{12}$ as a catalyst at 80 °C as well as selectively and in high yields leads to formamides of type **76**.[54]

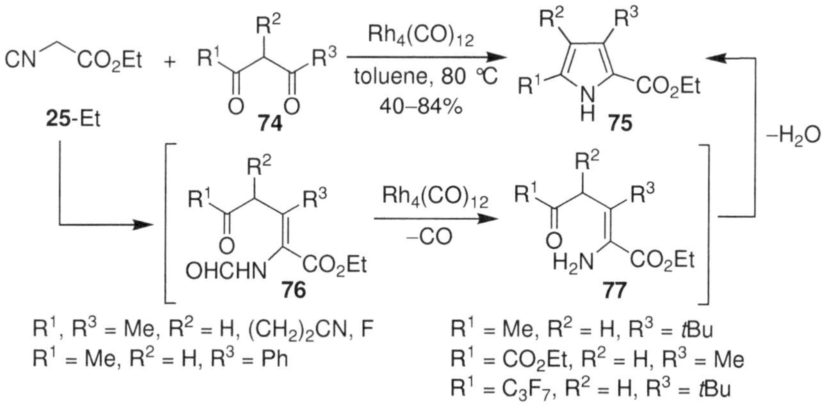

Scheme 19. Synthesis of tetrasubstituted pyrroles **75** by a rhodium-catalyzed reaction of ethyl isocyanoacetate (**25**-Et) with 1,3-dicarbonyl compounds **74**.[54]

When 1,3-dicarbonyl compounds are used as substrates in the reaction with **25**-Et, the rhodium-catalyzed decarbonylation of initially formed **76** was observed, and the amine **77** was formed, which is well set up to undergo cyclizing condensation to give the corresponding pyrrole **75**. The cocyclocondensation of **25**-Et with non-symmetric 1,3-dicarbonyl compounds ($R^1 \neq R^3$) leads to the corresponding pyrroles regioselectively when the substituents with essentially different steric or electronic demands were used.

2.2. α-Metallated *ortho*-Methylphenyl Isocyanides

The second type of metallated isocyanides, widely used in organic synthesis, are substituted *ortho*-methylphenyl isocyanides. Ito, Saegusa et al. first achieved the smooth deprotonation of *o*-methylphenyl isocyanides **78** by means of lithium dialkylamides in diglyme and utilized the thus obtained lithiated isocyanides **79** in versatile syntheses of various substituted indoles (Scheme

20).[56] When the reaction was carried out in THF or Et$_2$O, the addition of lithium dialkylamide onto the isocyano group became a competing process, decreasing the yield of indoles. An unsubstituted methyl group is lithiated selectively in the presence of a substituted one. *o*-Methylphenyl isocyanides with R^2 = H afforded the respective 3-unsubstituted indoles in high yields (82–100%) when lithium diisopropylamide (LDA) was used as a base, whereas for isocyanides substituted at the benzylic positions, lithium 2,2,6,6-tetramethylpiperidide (LiTMP) was the base of choice to provide 3-substituted indoles in good yields (62–95%).

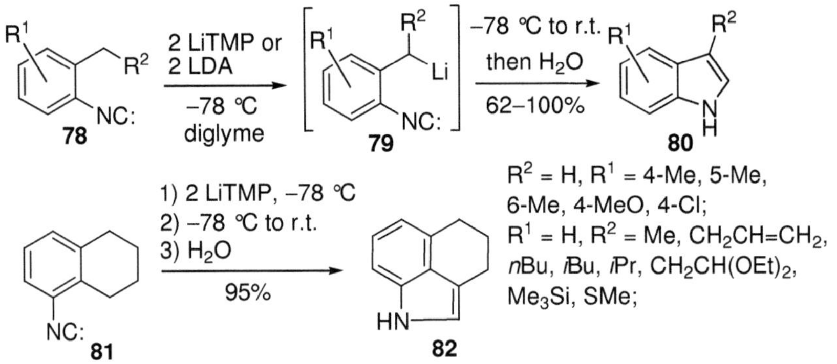

Scheme 20. Synthesis of indoles via lithiated *o*-methylphenyl isocyanides **79**.[56]

Using an excess of the base (2 equiv.) dramatically improved the yields of indoles which suggest, that the lithiation must be a reversible process. The tricyclic 1,3,4,5-tetrahydrobenz[c,d]indole **82** was obtained when 5,6,7,8-tetrahydronaphthalen-1-yl isocyanide **81** was used as a starting material. Different sequential reactions including the in situ modification of the *o*-methylphenyl isocyanides and employing different electrophiles have also been reported by the same authors. Thus, the cyclization of **79** at temperatures below −25 °C followed by trapping of the reaction mixture with various electrophiles such as alkyl halides, acid chlorides trimethylsilyl chloride and epoxides provides *N*-substituted indoles **85** exclusively in moderate to good yields (Scheme 21).[56b]

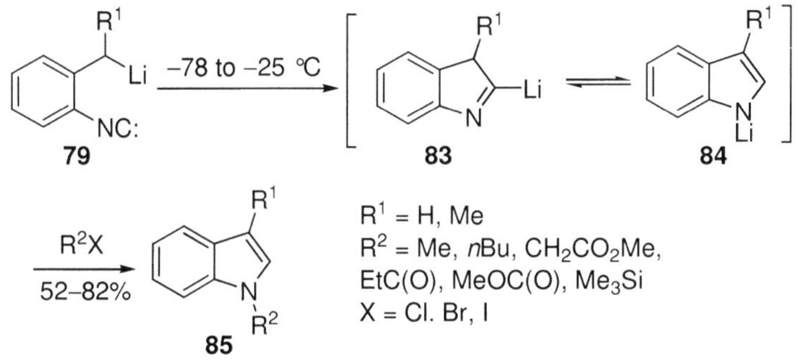

Scheme 21. Synthesis of 1,3-disubstituted indoles **85**.[56b]

Ito, Saegusa et al. reported, that acceptor-substituted *o*-methylphenyl isocyanides can be conveniently converted into the corresponding 3-substituted indoles under Cu(I) catalysis (Scheme 22).[57, 58]

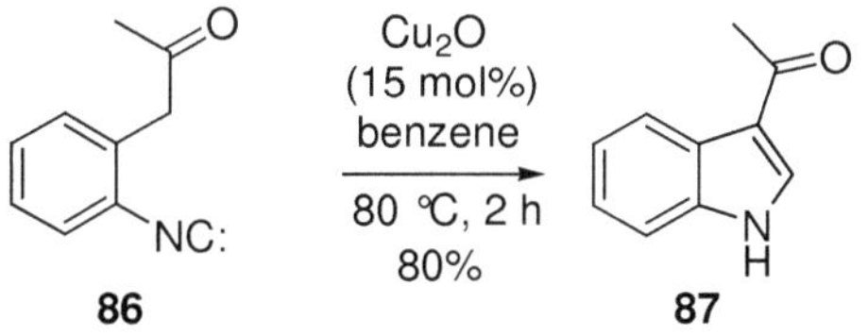

Scheme 22. Cu$_2$O-catalyzed synthesis of 3-acylindole **87**.[57,58]

This method usefully supplements the approach to substituted indoles via lithiated *o*-methylphenyl isocyanides (vide supra). Thus, in the Cu$_2$O-catalyzed reaction some functional groups, such as keto carbonyl groups are tolerated (3-acylindoles of type **87**, for example, could not be prepared by means of benzylic lithiation)[58] while the base-mediated variant does not require acceptor substituents in the side chain of the aryl isocyanide.[56] The key intermediate of this process is supposed to be an α-copper-substituted (acylmethyl) phenyl isocyanide, which undergoes an intramolecular insertion of the isocyano group into the newly formed C-Cu bond to provide, after isomerization and protonation, indoles of type **87**. The evidences for intermolecular insertions of isocyanides into copper(I) complexes of "active hydrogen" compounds like acetylacetone, malonates and others[59] support this assumption.

α,α-Disubstituted *o*-methylphenyl isocyanides of type **88** in turn furnished the respective 3,3-disubstituted-3*H*-indoles **89** in moderate to high yields (Scheme 23).[57]

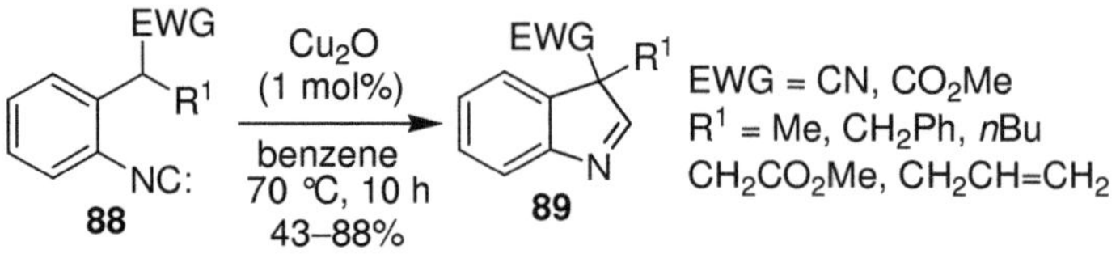

Scheme 23. Synthesis of 3,3-disubstituted 3*H*-indoles **89**.[57]

Various substituted *o*-methylphenyl isocyanides could be prepared by alkylation of *o*-(lithiomethyl)phenyl isocyanides with alkyl halides and reactions with other electrophiles, such as epoxides, trimethylsilyl chloride, dimethyl disulfide,[56b] aldehydes (ketones),[60] isocyanates and isothiocyanates, respectively.[61] The corresponding adducts may be involved in subsequent base-promoted or Cu(I)-catalyzed cyclizations to furnish indoles and other benzoannelated heterocycles. Thus, adducts of type **90** of reaction of *o*-(lithiomethyl)phenyl isocyanide (**97**) with isocyanates can undergo two types of Cu$_2$O- catalyzed cyclizations providing 3-substituted indoles **91**, benzodiazepine-4-ones **92** or both of them depending on the substituents present (Scheme 24), while in a base-mediated

cyclization of *N*-substituted *o*-(isocyanophenyl)acetamides **90** (and analogous thioacetamides), indoles of type **91** are obtained exclusively.[61]

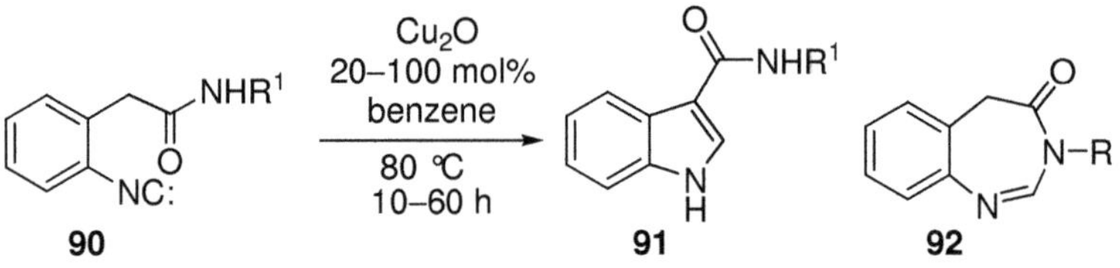

R	91 (%)	92 (%)
n-C$_4$H$_9$	0	85
c-C$_6$H$_{11}$	25	58
t-C$_4$H$_9$	20	0
Ph	75	0

Scheme 24. Cu$_2$O-catalyzed cyclizations of *N*-substituted *o*-isocyanophenylacetamides **90**.[61]

The reaction of *o*-(lithiomethyl)phenyl isocyanides **79** with aldehydes (ketones) at −78 °C, hydrolysis of the reaction mixture at the same temperature and subsequent Cu$_2$O-catalyzed cyclization of the respective isocyanoalcohols **93** prepared in this way, furnishes 4,5-dihydro-3,1-benzoxazepines **94** in high overall yields. An analogous cyclization of the adduct **95** of *o*-

(lithiomethyl)phenyl isocyanide (**97**) with 1-butene epoxide leads to 4*H*-5,6-dihydro-3,1-benzoxacine **96** in 42% yield (Scheme 25).[60]

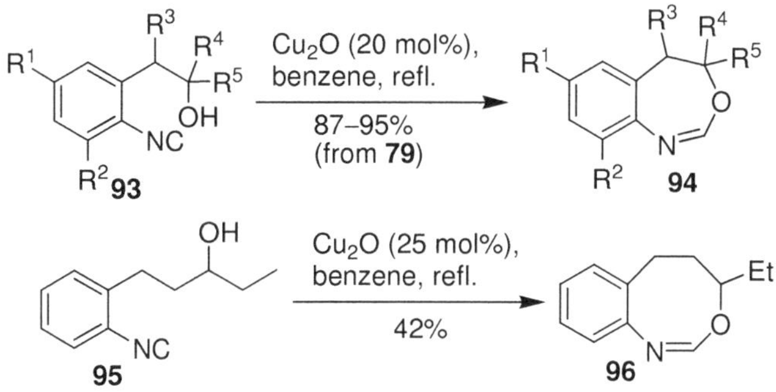

Scheme 25. Synthesis of 4,5-dihydro-3,1-benzoxazepines **94** and 4*H*-5,6-dihydro-3,1-benzoxacine **96**.[60]

Substituted *o*-methylphenyl isocyanides prepared by functionalization of *o*-(lithiomethyl)phenyl isocyanide **97** can undergo hydrolysis to provide anilines, and subsequent cyclization of the latter by the reaction with an adjacent keto or ester group provides 2-substituted indoles **99**[58] or 1,3,4,5-tetrahydro-2*H*-benzazepine-2-ones **101**, respectively (Scheme 26).[62] These representative examples show applications of isocyanides as *masked amines*.

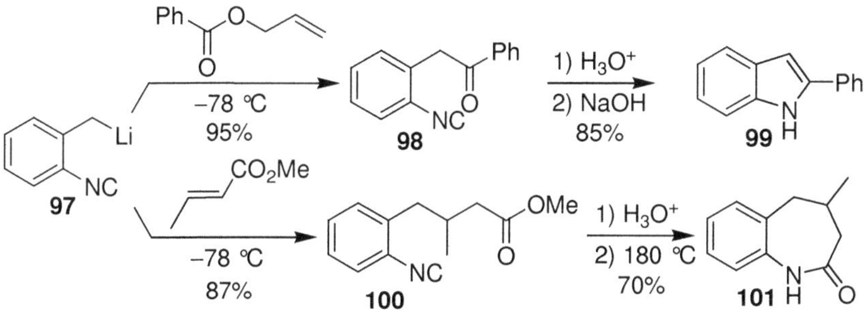

Scheme 26. Synthesis of indole **99** and cyclic amide **101** from **97**.[58,52]

On the other hand, the adducts of **79** with aldehydes (ketones), isocyanoalcohols of type **93**, have been reported to undergo a further Lewis-acid catalyzed rearrangement to *N*-formylindolines **103** (Scheme 27).[63]

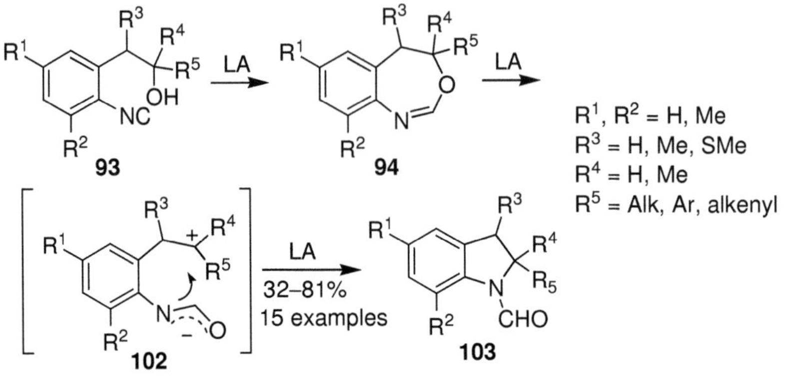

Scheme 27. Synthesis of *N*-formylindolines **103** by Lewis-acid catalyzed isomerization of isocyanoalcohol **93**.[63]

The reaction is supposed to proceed with initial formation of the dihydro-3,1-benzoxazepines **94** by Lewis acid-catalyzed insertion of the isocyano group into the O-H linkage. This initial product undergoes heterolytic cycloreversion and re-cyclization of the zwitterionic intermediate of type **102** to yield the *N*-formylindolines **103**. Dihydro-3,1-benzoxazepines **94** prepared independently, in turn undergo the same Lewis-acid catalyzed rearrangement to provide **103**.[63]

An interesting precedent of a catalytic C, H-activation on 2,6-dimethylphenyl isocyanide (**104**) and some other similar aryl isocyanides by ruthenium complexes **106** and **107** leading to indoles **105** has been reported by Jones et al.[64] along with interesting mechanistic investigations of this transformation.[64b] Unfortunately, this method implies harsh reaction conditions (140 °C, 94 h) and has only a very limited scope. Moreover, the thermal instability of *o*-methylphenyl isocyanides as well as (reversible) insertion of isocyanide into the N-H bond of the newly formed indole molecule decreases the yields of final products and prolongs the reaction times.[64]

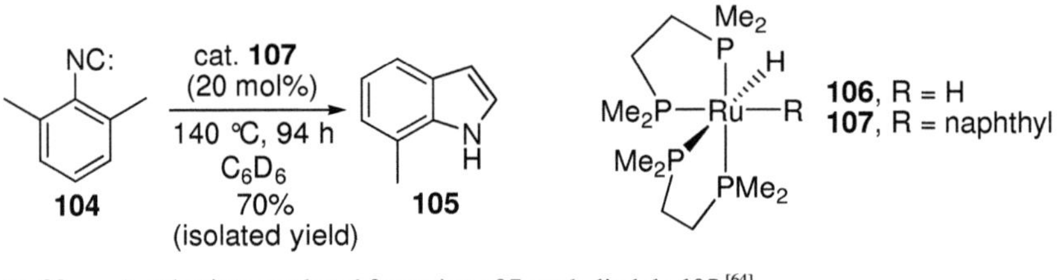

Scheme 28. A ruthenium-catalyzed formation of 7-methylindole **105**.[64]

2.3. Other Metallated Isocyanides

Kobayashi et al. have reported on the synthesis of 4-hydroxyquinolines **110** by a magnesium bis(diisopropylamide)-induced cyclization of keto ester (or keto amide) **109**. The latter is generated in situ by a Claisen-type condensation of *ortho*-isocyanobenzoate **108** with magnesium enolates of alkyl acetates or N,N-dimethylacetamide (Scheme 29).[65]

On the other hand, 2-(2-isocyanophenyl)acetaldehyde dimethyl acetals of type **111** upon treatment with an excess of LDA at −78 °C in diglyme furnish 3-methoxyquinolines **112** in good to high yields (Scheme 30).[66] The intermediate lithiated isocyanide **114** is believed to arise by deprotonation of **111** at the benzylic position, subsequent elimination of lithium methoxide to give the corresponding *o*-isocyano-β-methoxystyrene **113** followed by lithiation of the latter at the β-position.

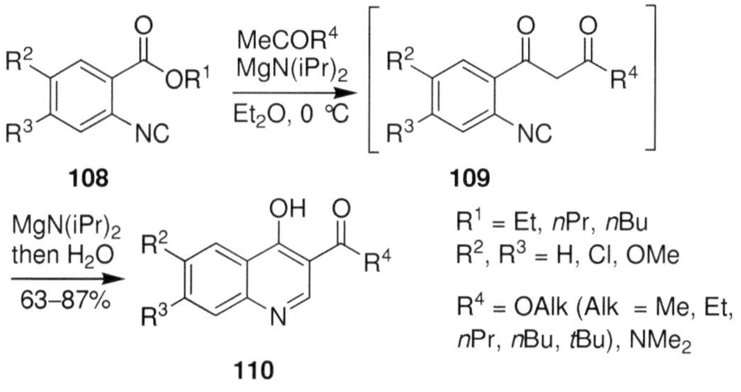

Scheme 29. Synthesis of 4-hydroxy-3-quinolinecarboxylic acid derivatives **110**.[65]

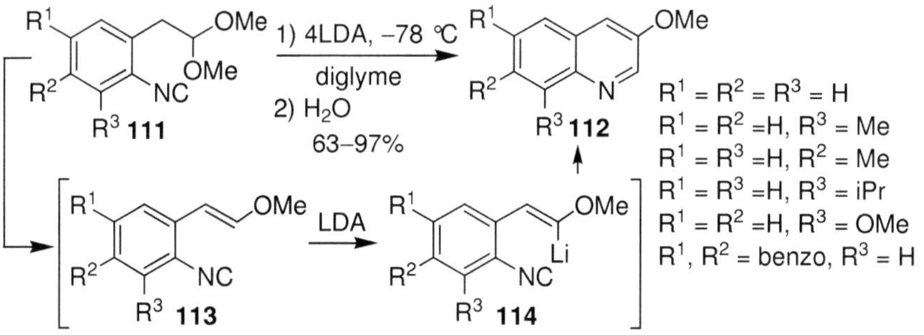

Scheme 30. Synthesis of 3-methoxyquinolines **112**.[66]

3. Addition to the Isocyano Group Followed by a Cyclization

3.1. Non-Catalyzed Processes.

Organolithium[67] as well as organomagnesium[68] reagents have been shown to undergo α-addition to isocyanides to provide metalloaldimines, which can undergo cyclizations to give the corresponding *N*-heterocycles if there is an appropriate adjacent functional group.

Thus, the addition of *t*BuLi to phenyl isocyanide (**115**) followed by a directed *ortho*-lithiation assisted by TMEDA has been reported to lead to the formation of the dilithiated aldimine **116**, which in turn can be trapped with various elementchlorides to provide various benzazoles **117** in moderate yields (Scheme 31).[69]

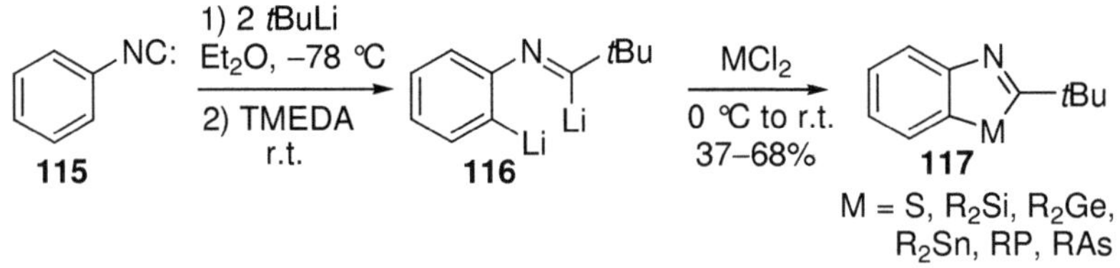

Scheme 31. Addition of *t*BuLi/*ortho*-lithiation of phenyl isocyanide (**115**). Synthesis of benzoannelated azoles **117**.[69]

Using an excess of the bulky *t*BuLi (2 equiv.) and adding the isocyanide to the organolithium reagent has been found to be crucial for the effective formation of **117**. The resulting conventional benzazoles (benzothiazoles) as well as some unusual benzazoles (e. g. benzoazosiloles, benzoazogermoles etc.) have been investigated and compared from the viewpoint of their possible aromaticity.[69]

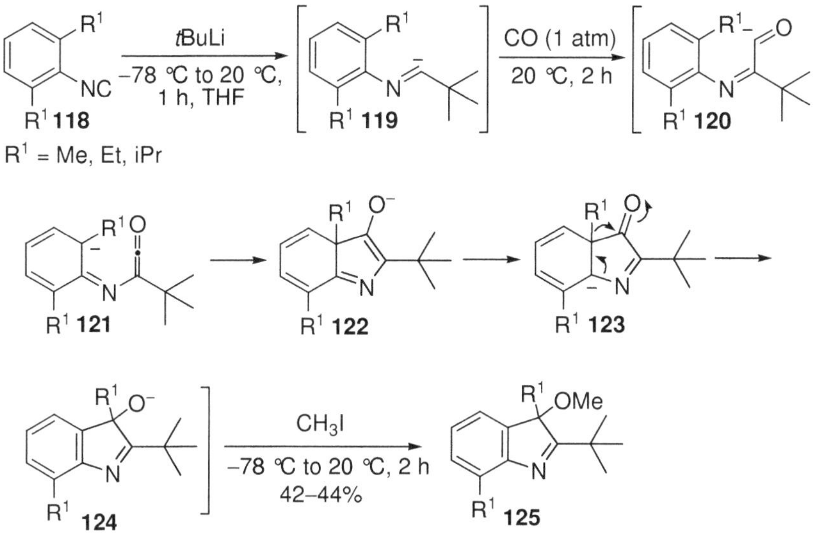

Scheme 32. Synthesis of 3H-indoles **125**.[70]

To avoid possible *ortho*-metallation after the addition of *t*BuLi onto the isocyano group, Murai et al. have used 2,6-dialkylphenyl isocyanides **118**. The resulting deprotonated aldimines **119** have been trapped with carbon monoxide to induce a complicated cascade of transformations leading, after treatment with methyl iodide, to 3H-indoles **125**.[70] The proposed mechanism starts with the formation of the aforementioned lithioaldimine **119**, which is transformed to the reactive acyllithium intermediate **120**, upon treatment with CO. The formation of the non-aromatic ketene **121** followed by a cyclization to alcoholate **122**, its tautomerization to the ketone **123** and final alkyl group migration afford the deprotonated 3H-indole **124**, which reacts with methyl iodide to finally give the isolated 3-methoxy-3H-indole **125** (Scheme 32).[70]

A convenient and efficient synthesis of 2,3-disubstituted quinolines **127** by the reaction of nucleophiles such as alcohols, amines and sodium enolate of diethylmalonate with *ortho*-alkynylphenyl isocyanides **126** has been reported by Ito et al. (Scheme 33).[71] A related diethylamine-induced 6-*endo-dig* cyclization of *o*-isocyanobenzonitrile **128** afforded 2-diethylaminoquinazoline **129** in quantitative yield (Scheme 33).

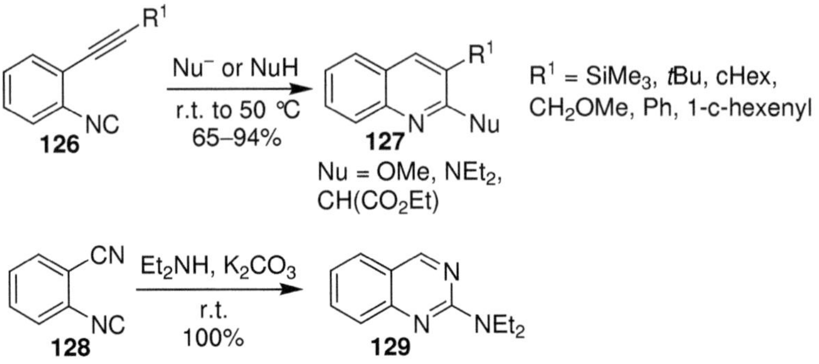

Scheme 33. Synthesis of 2,3-disubstituted quinolines **127** and 2-diethylamino-quinazoline **129**.[71]

In the crucial step of both of these processes, the imidoyl anion, initially formed after the addition of a nucleophile onto the isocyano group, is supposed to undergo a 6π-electrocyclization, subsequent isomerization and protonation to give **127** or **129**.[71]

Known reactions of other potential precursors of heterocycles, 1,2-diisocyanoarenes **130**, with nucleophiles are limited to that with Grignard reagents. Quinoxaline oligo- and polymers **131** (Scheme 34) with different order of polymerization depending on the substituents, isolated after hydrolysis of the reaction mixture of such 1,2-diisocyanoarenes with alkylmagnesium bromides, apparently arise from successive insertion of isocyano groups into magnesium-carbon bonds.[72]

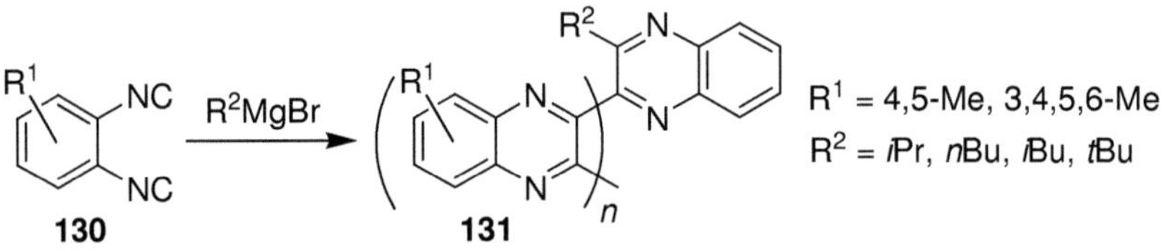

Scheme 34. Oligomerization of 1,2-diisocyanoarenes (**130**) by treatment with Grignard reagents.[72]

Kobayashi et al. have shown, that *o*-isocyano-β-methoxystyrenes such as **132** can be employed in the synthesis of 2,4-disubstituted quinolines **135** (Scheme 35).[73] Organolithium reagents, lithium dialkylamides and lithium thiophenolate undergo an α-addition onto the isocyano group to provide the imidoyl anion **133**, which after a cyclization and subsequent elimination of methoxide, gives quinolines **135** in low to high yields.

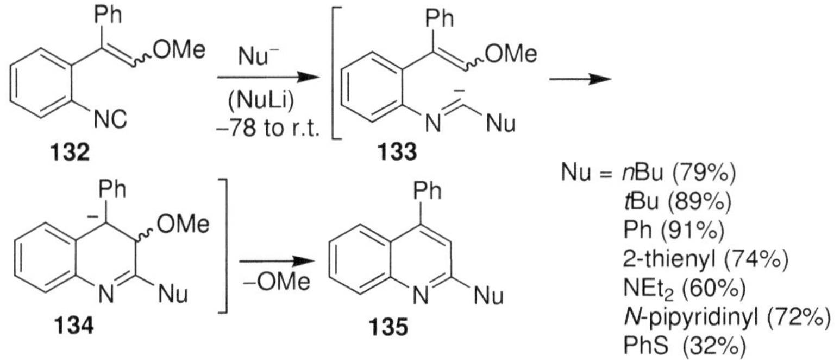

Scheme 35. Synthesis of quinolines **135** by addition of nucleophiles to *o*-isocyano β-methoxystyrenes **132**.[73]

Independently, Ichikawa et al. have reported on a similar reaction of organometallic reagents with β,β-difluoro-*o*-isocyanostyrene (**136**) leading to 2-substituted 3-fluoroquinolines (**137**) by 6-*endo*-trig cyclization of initially formed acyllithium

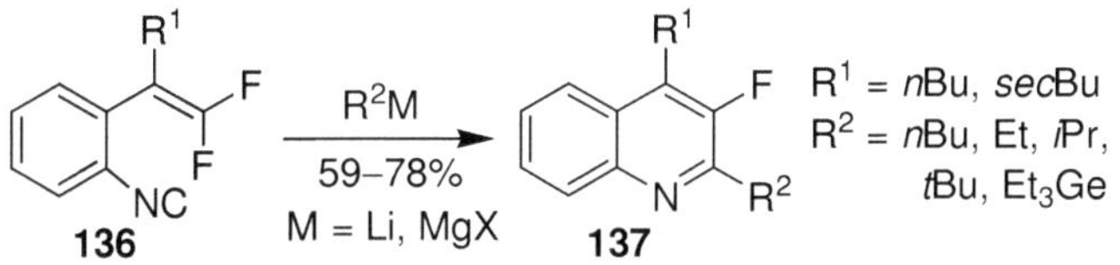

Scheme 36. Synthesis of 2,3,4-trisubstituted quinolines **137** by reaction of organometallic reagents with β,β-difluoro-*o*-isocyanostyrenes **136**.[74,75]

intermediates with elimination of a fluoride anion (Scheme 36).[74] *n*BuLi reacts to give a complicated mixture of products, whereas sterically encumbered *t*BuLi leads to the formation of the corresponding quinoline in 78% yield. Alkylmagnesium reagents less reactive than alkyllithiums, have also been successfully employed in this reaction as well as triethylgermyl- and tributylstannyllithium.[75]

Isocyanides have been known to react with acyl halides to provide the corresponding α-keto imidoylhalides. The products of these insertions such as **139** derived from 2-phenylethyl isocyanides of type **138** have been reported to undergo subsequent Ag(I)-mediated cyclizations to form 1-acyl-3,4-dihydroisoquinolines **141** in moderate to good yields (Scheme 37).[76] The authors suggest, though without any evidence, that transient acylnitrilium cations of type **140** are intermediates in these reactions under ionizing conditions (Ag salts) while in the presence of Lewis

(SnCl$_4$) or Brønsted acids (CF$_3$SO$_3$H), the corresponding protonated or coordinated halo iminium derivatives **139** play the same role. Apart from dihydroisoquinolines obtained by this method, the furan- and indole-annelated products of type **142** and **143** (Scheme 37) have been synthesized in the same manner in good yields. The generality of this method and very mild conditions make it a useful supplement to the classical Bischler-Napiralski synthesis of 3,4-dihydroisoquinolines and the respective isoquinolines. Compound **144** with the skeleton of the alkaloid erythrinane has been also conveniently prepared in a two-step one-pot procedure from the 3,4-dihydroisoquinoline of type **141**.[76b]

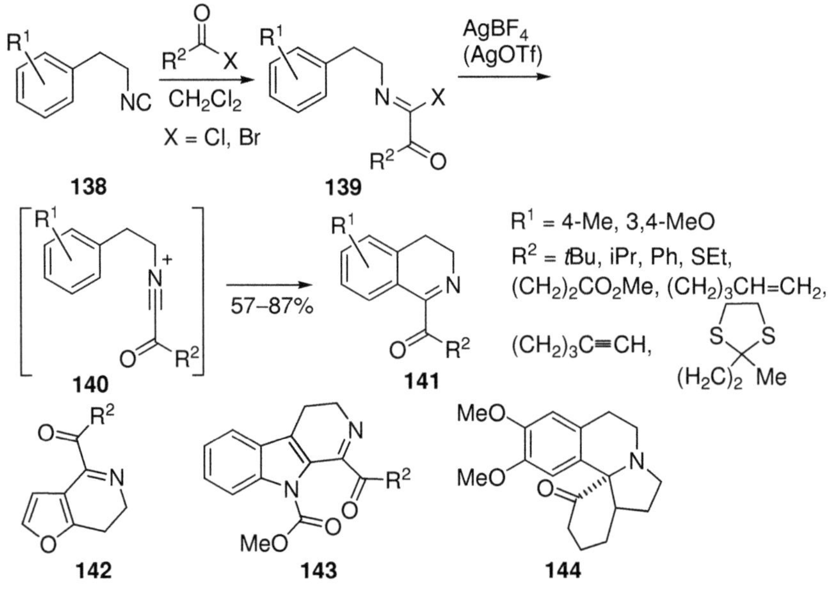

Scheme 37. Ag(I)-mediated cyclization of 2-ethylphenyl isocyanides **138**.[76]

3,4-Donor-disubstituted 3-phenylpropyl isocyanides of type **145** which are homologous to the previously discussed isocyanides **138**, also smoothly undergo addition of an acid chloride with subsequent Ag(I)-promoted cyclization to furnish 2-acylbenzazepines **146** (Scheme 38, eq. (1)).[77] However, similar isocyanides of type **147** with another substi-tution pattern, instead of forming products of type **146**, tend to undergo spiroannelation of the corresponding intermediate imidoyl chlorides to give, after in situ desilylation and tautomerization, spirocyclic Δ^1-piperidienes **148** exclusively in good yields (Scheme 38, eq. (2)).[77, 78]

Similarly, the addition of the acyl chloride **150** onto the isocyano group of the isocyano silylenolate **149**, subsequent AgBF$_4$-mediated cyclization of the corresponding intermediate and final deprotection of the *tert*-butyldimethylsilyl ether provide the 2-acyl Δ^2–pyrroline **151**, a key intermediate in a total synthesis of the alkaloid (±)-dendrobine (**152**) reported by Livingouse et al. (Scheme 39).[79] Some other unactivated alkenes have later also been employed in this Ag(I)-mediated cyclization to provide the respective 3,4-dihydro-2*H*-pyrroles or 3,4,5,6-tetrahydropyridines in moderate to good yields.[80, 81]

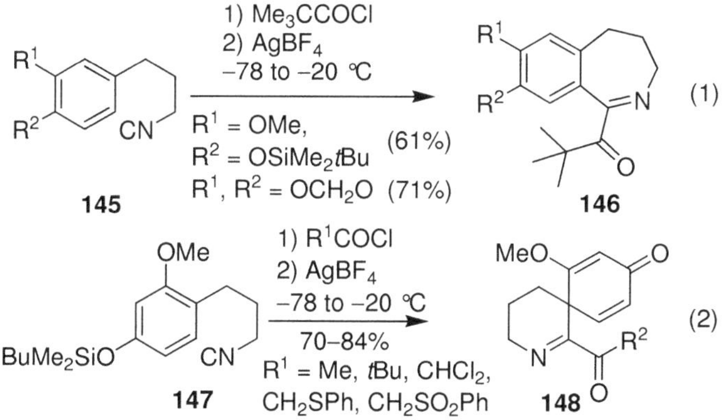

Scheme 38. Cyclizations of arylisocyanopropanes **145** and **147**.[77, 78]

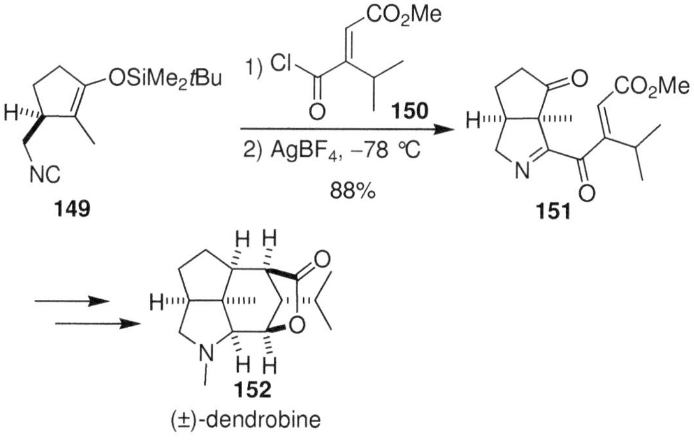

Scheme 39. Synthesis of 2-acyl Δ^2–pyrroline **151**, an intermediate in the total synthesis of (±)-dendrobine **152**.[79]

Similarly to acid chlorides, arylsulfenyl chlorides (ArSCl) react with isocyanides leading to unstable *N*-alkoxycarbonyl-*S*-arylisothiocarbamoyl chlorides such as **153**, which are capable of further cyclization if an appropriate adjacent functionality is present. Thus, the adducts of isocyanides **156** with ester or amide moieties have been shown to undergo subsequent cyclizations to 2-arylthio-5-alkoxyoxazoles **154**[82] and 3-alkyl-2-arylthio-1,3-diazolium-4-olates **155**,[83] respectively, upon treatment with triethylamine (Scheme 40). Similarly, dichlorosulfane SCl$_2$ reacts with two equivalents of the isocyanide **156** to provide, after amine-induced cyclization, the corresponding 2,2'-bis(oxazolyl)sulfide.[84] The reaction of ethyl isocyanoacetate with dichlorodisulfane S$_2$Cl$_2$ unexpectedly led to the formation of thiazolo[5,4-d]thiazole-2,5-dicarboxylate **157** (Scheme 40).[84] The mechanism of this complex transformation proposed by the authors (not presented here) includes cleavage of the S-S bond at an early stage followed by a cascade of further transformations.[85]

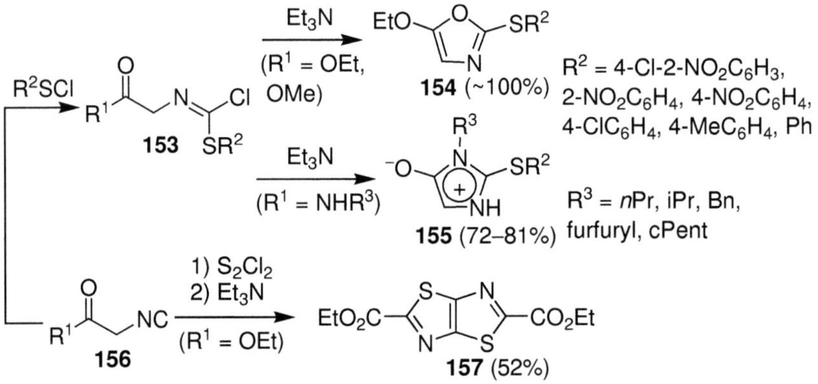

Scheme 40. Reactions of isocyanides **156** with arylsulfenyl chlorides and dichlorodisulfane followed by Et$_3$N-induced cyclizations.[82, 83, 84]

3.2. Transition Metal-Catalyzed Processes

Aryl isocyanides have been shown to react with selenium to form isoselenocyanates.[86] The same reaction with alkyl isocyanides in the presence of a base followed by reactions of these isoselenocyanates with amines or alcohols to give selenoureas or selenocarbamates, respectively, has also been reported.[87] With *o*-halophenyl isocyanides **159** as substrates in this reaction, the resulting selenocarbamates **158** have been efficiently transformed into the corresponding benzoselenazoles **160** in a CuI-catalyzed one-pot process.[88] Secondary alkyl- and arylamines, *n*-

butylamine as well as imidazole were converted into the respective 2-substituted benzoselenazoles **160** in high yields (Scheme 41, eq. (1)).

When alcohols or thiols were used instead of amines, 2-oxy- or 2-thiabenzoselenazoles (**161**) were obtained in high yields under essentially the same conditions as previously, but without a base. Aliphatic alcohols and phenols with electron-donating substituents gave remarkably higher yields than 4-methoxycarbonylphenol (48%), while all tested thiols, both aromatic and aliphatic, provided the corresponding products **161** in high yields (Scheme 41, eq. (2)).[88]

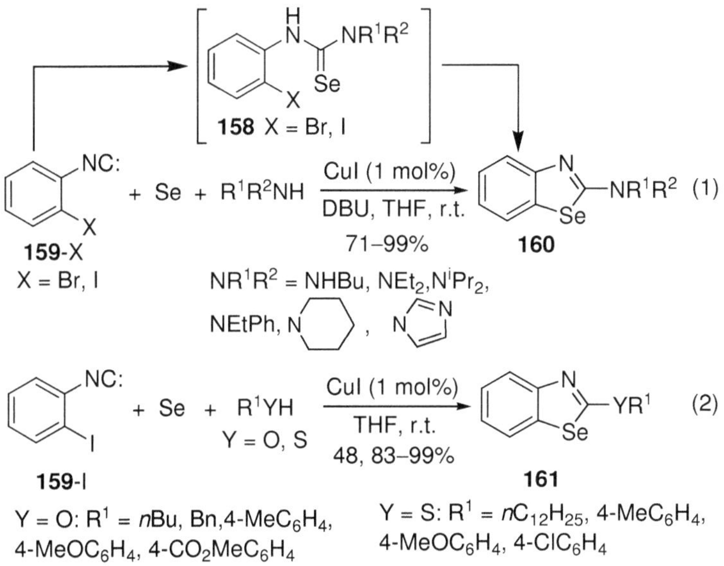

Scheme 41. Copper(I)-catalyzed synthesis of benzoselenazoles **160** and **161**.[88]

Further investigations revealed, that *ortho*-bromophenyl isocyanide (**159**-Br) can form 2-aminobenzoselenazoles **160** even without a copper catalyst, though the reaction proceeds more slowly and only at elevated temperature (100 °C), but **160** wasformed even at ambient temperature from the *ortho*-iodophenyl isocyanide **159**-I. This led the authors to propose that mechanistically, the cyclization might proceed through an intramolecular nucleophilic aromatic substitution of the initially formed selenolate **162** via an intermediate of type **163**. As the role of copper in facilitating this process remains unclear, the possibility of an alternative route including a copper-catalyzed cross-coupling reaction via intermediates of type **164** should not be ruled out either.[89]

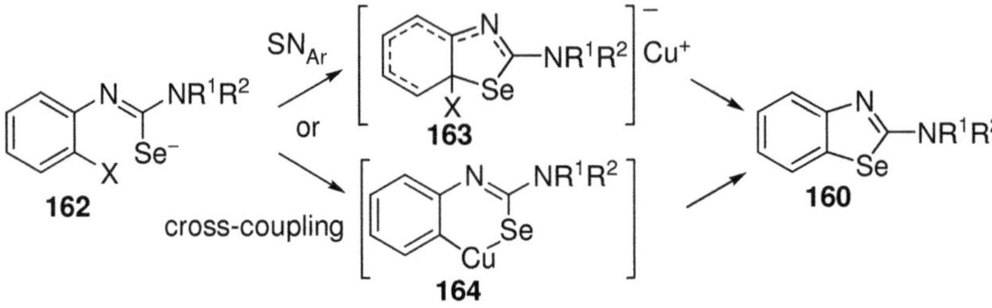

Scheme 42. Proposed mechanisms for the formation of benzoselenazoles **160**.[89]

The same authors have also extended their earlier developed tellurium-assisted imidoylation of amines with isocyanides[90] and used the thus formed intermediates **165** in a copper(I)-catalyzed one-pot synthesis of 2-amino-1,3-benzotellurazoles **166** (Scheme 43).[88]

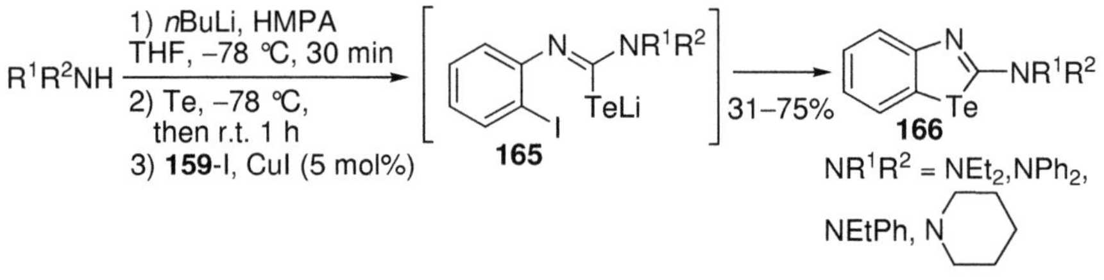

Scheme 43. The synthesis of benzotellurazoles **166**.[88]

4. Goals of this Study

Critical analysis of the relevant literature revealed, that although a plethora of useful processes, for the construction of nitrogen heterocycles from isocyanides are known, there are still a lot of gaps to be filled. Particularly, the metal-catalyzed processes still remain limited, although evident interest of researchers has recently been devoted to this topic. Thus, three different directions of research work have been chosen after the first promising experimental tests confirming some theoretical suppositions:

1) Synthesis of substituted pyrroles:
- Further elaboration of a pyrrole synthesis from substituted methyl isocyanides and acetylenes previously developed in our group. An extensive study of the scope and the limitations of this method as well as an investigation of the mechanism of the copper-catalyzed reaction
- Extension of this method to the use of terminal *unactivated* acetylenes for the synthesis of pyrroles

2) Chemistry of ortho-metallated aryl isocyanides:
- Development of methods for generating *ortho*-metallated phenyl isocyanide and its heteroanalogues as novel building blocks with a potentially broad scope of synthetic applications
- Investigation of the reactions of *ortho*-lithiophenyl isocyanide with various isocyanates (isothiocyanates) focusing on the synthesis of pharmaceutically relevant heterocycles and naturally occurring alkaloids
- Investigation of the reactions of *ortho*-lithiophenyl isocyanide with various carbonyl compounds focusing on the synthesis of heterocycles

3) Synthesis of benzimidazoles:
- Development of an efficient (catalytic) approach to substituted benzimidazoles and related heterocycles by the reaction of *ortho*-haloaryl isocyanides with primary amines

B. Main Part

1. Oligosubstituted Pyrroles Directly from Substituted Methyl Isocyanides and Acetylenes[91]

Background and Preliminary Considerations

Oligofunctional pyrroles play a pivotal role among five-membered heterocycles, being basic constituents of numerous natural products,[92] potent pharmaceuticals,[93] molecular sensors and other devices.[94] Therefore, considerable attention has been paid to develop efficient general methods for the synthesis of pyrroles,[95] and in recent years, their amount has been significantly extended.[96] Some methods of pyrrole synthesis such as Barton-Zard[16] and van Leussen[32] syntheses (vide supra) are based on isocyanides. Among them, the addition of isocyanides **63** onto the triple bond of acetylenes **64** developed by Yamamoto et al.[50] and de Meijere et al.[51] is the most direct and therefore one of the most promising (see Scheme 16 in Introduction).

Various 2,3,4-trisubstituted pyrroles **65**, bearing sulfonyl, dialkoxyphosphoryl, trifluoromethyl, cyano and secondary amino groups have been previously prepared in one step from readily available acetylenes and acceptor-substituted methyl isocyanides.[51]

As a continuation of the work started in our group by O. Larionov, the scope and limitations of this pyrrole synthesis were systematically investigated focusing on the copper-catalyzed variant. Terminal acceptor-substituted acetylenes have not been employed in this reaction so far as well as both teminal and internal *unactivated* acetylenes. This provided a challenge for a further development of this direct pyrrole synthesis.

Synthesis of 2,3,4-Trisubstituted and 2,4-Disubstituted Pyrroles

In this study, particular attention has been paid to aryl and hetaryl-substituted acceptor acetylenes **64**, as only a few examples of such compounds have been employed earlier. The substituted methyl propiolates **168** have been routinely prepared from the corresponding terminal acetylenes **167** by lithiation with *n*BuLi followed by a reaction with methyl chloroformate (Scheme 44). The thienyl-derivative **168g** was prepared according to a literature procedure from thiophene and tetrachlorocyclopropane **169** (Scheme 44). This interesting one-pot transformation furnished **168g** in 95% overall yield.

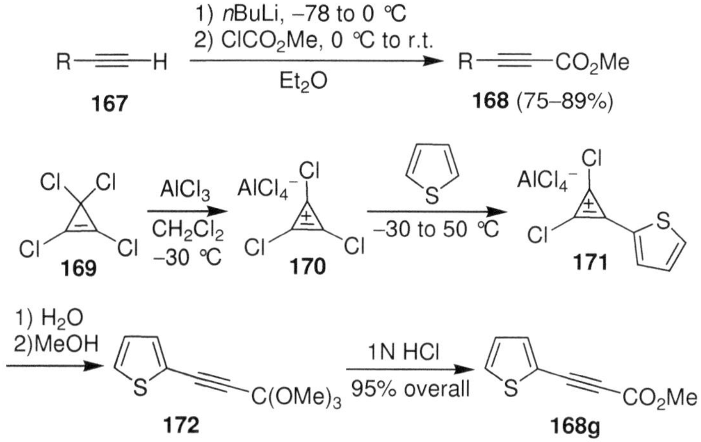

Scheme 44. Synthesis of substituted methyl propiolates **168**.

With thus prepared methyl propiolates **168**, various new pyrroles have been synthesized (Table 1). Both CuSPh-catalyzed (at 85 °C) and KO*t*Bu-mediated reaction of methyl cyclopropylpropiolate (**168a**) with ethyl isocyanoacetate (**25**-Et) proceeded smoothly to provide the corresponding pyrrole **173ba** in 87 and 89%, respectively. Methyl (1-methoxyethyl) propiolate **168b** afforded the pyrrole **173ab** only in 54% yield in a CuSPh-catalyzed reaction with **25**-Me in the same conditions, although the conversion of starting materials was quantitative according to NMR data of the reaction mixture. The dimerization of isocyanide **25**-Me to form imidazole of type **174** is most likely the main side-reaction in this case. Aryl- (hetaryl-) substituents in acetylenes **168** were expected to cause reduced reactivity towards nucleophiles. It was reported earlier,[51] that the reaction of ethyl phenylpropiolate with *tert*-butyl cyclopropylpropiolate was quite sluggish at 80 °C, but readily furnished the corresponding pyrrole at higher temperature (120 °C). Other acetylenes with aryl- (**168c, 168d, 168e**) and heteroaryl substituents (**168f, 168g**) were also efficiently converted into the respective pyrroles **173ac–173ag** using this method (entries 3–7).

Table 1. Various 2,3,4-trisubstituted pyrroles **173** prepared by the formal cycloaddition of substituted methyl isocyanides **63** to acetylenes **168**.[97]

Entry	Isocyanide	Acetylene	Product	Method,[a],[b] Yield (%)[c]
1	CN-CH₂-CO₂Et **25-Et**	**168a** (cyclopropyl-C≡C-CO₂Me)	**173ba**	A, 89; B,[d] 87
2	CN-CH₂-CO₂Me **25-Me**	**168b** (CH(OMe)Me-C≡C-CO₂Me)	**173ab**	B,[d] 54
3	CN-CH₂-CO₂Me **25-Me**	**168c** (4-EtO-C₆H₄-C≡C-CO₂Me)	**173ac**	B, 75
4	CN-CH₂-CO₂Me **25-Me**	**168d** (4-F-C₆H₄-C≡C-CO₂Me)	**173ad**	B, 78
5	CN-CH₂-CO₂Me **25-Me**	**168e** (4-F₃C-C₆H₄-C≡C-CO₂Me)	**173ae**	B, 70

Table 1 (continued). Various 2,3,4-trisubstituted pyrroles **173** prepared by the formal cycloaddition of substituted methyl isocyanides **63** to acetylenes **168**.

Entry	Isocyanide	Acetylene	Product	Method,[a],[b] Yield (%)[c]
6	CN–CH(CO₂Me) **25-Me**	2-pyridyl-C≡C-CO₂Me **168f**	**173af** (2-Py, CO₂Me, MeO₂C, NH)	B, 68
7	CN–CH(CO₂Me) **25-Me**	2-thienyl-C≡C-CO₂Me **168g**	**173ag**	B, 94
8	CN–CH(CO₂Et) **25-Et**	H-C≡C-CO₂Me **168h**	**173bh**	A, 35; B,[d] 37
9	CN–CH(SO₂Tol) **41-H**	H-C≡C-CO₂Me **168h**	**173ch**	A, 38; B,[d] 30
10	CN–CH₂Ph **63e**	H-C≡C-CO₂Me **168h**	**173eh**	A, 7; B,[d] 25
11	CN–CH₂–C₆H₄–NO₂ **63f**	H-C≡C-CO₂Me **168h**	**173fh**	B,[d] 44

[a] Method A: Addition of KOtBu (1.2 equiv.), 1 h, then 1 h at 20 °C, THF. [b] Method B: CuSPh (5 mol %), DMF, 120 °C, 12 h. [c] Yield of isolated product. [d] The reaction was carried out at 85 °C. [e] The reaction was carried out at 60 °C.

With electron-acceptor substituted terminal acetylenes, the yields of pyrroles **173** were dramatically lower. Thus, the CuSPh-catalyzed reactions of isocyanides **63** with methyl propiolate **168h** led to the corresponding pyrroles **173** in only 25–44% yields, and in the presence of KOtBu the yields of pyrroles **173** were also low (Table 1, entries 8–11). It is known, that methyl propiolate **168h** easily dimerizes to give dimethyl hex-2-en-4-ynedioate both under base[98] and copper(I)[99] catalysis, complicating this reaction. Even with an excess of **168h**, the yields of pyrroles **173** were not any better.

The copper(I)-catalyzed variant of the reaction is of great interest, because it may bring along certain advantages in cost, efficiency and compatibility with base-sensitive substrates. Different solvents were tested in the copper-catalyzed cycloaddition of *p*-toluenesulfonylmethyl isocyanide (TosMIC, **41**-H) to methyl cyclopropylacetylene carboxylate (**168a**). Dimethylformamide (DMF) turned out to be the solvent of choice giving a better yield of pyrrole **173da** than any other tested solvent (toluene, ethanol, ethyl acetate, acetonitrile, 1,2-dichloroethane, dioxane).

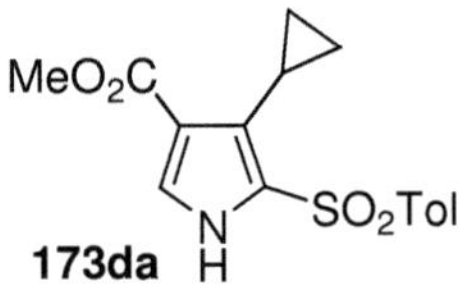

$$\text{MeO}_2\text{C} \quad \text{SO}_2\text{Tol}$$
173da

Figure 2. The pyrrole **173da**.

Among all copper catalysts tested, copper(I) thiophenolate and preactivated nanosize metallic copper powder in DMF at 90 °C turned out to be the most efficient, providing the pyrrole **173ba** in 93 and 92% yield, respectively (Table 2, entries 1 and 2). With copper(I) oxide as a catalyst, **173ba** was obtained in a slightly lower yield of 78%, whereas other copper compounds gave inferior results. Surprisingly, some copper(II) compounds (copper(II) acetylacetonate, copper(II) acetate, entries 12, 13, respectively) catalyzed the formation of **173ba** as well, and gave even better results than CuI and other copper(I) halides. Copper(II) compounds are supposed to be (fully or partly) reduced by isocyanides to the corresponding Cu(I) salts, which actually catalyze the reaction.

It is conceivable, that this reaction could be used for many other applications, for example, in biological systems, if it fulfilled the demands of Sharpless' so-called "click" chemistry,[100] i.e. provided good yields and could be carried out at low temperatures.

Table 2. Optimization of the copper catalyst for the synthesis of **173aa**.[a]

Entry	Cu catalyst	Yield of **173aa** (%)[b]
1	CuSPh	93
2	Cu-NP[c]	92
3	CuSePh	45
4	CuSHex	52
5	CuPPh$_2$	39
6	Cu$_2$O	78
7	Cu$_2$S	37
8	CuCl	6
9	CuBr	10
10	CuI	3
11	CuCN	24
12	Cu(acac)$_2$	24
13	Cu(OAc)$_2$·H$_2$O	19
14	CpCuP(OMe)$_3$	89[d]

[a] Reagents and conditions: **25**-Me (1.1 mmol), **168a** (1.0 mmol), copper salt (5 mol%), DMF (2 mL). [b] Determined by ^{1}H NMR with hexamethylbenzene as an internal standard. [c] The abbreviation Cu-NP stands for preactivated nanosize copper powder. [d] The reaction was carried out at 20 °C with 10 mol% of catalyst and, according to TLC, was completed within 2 h.

With the intention to achieve this kind of pyrrole formation at temperatures lower than 70 °C (the lowest temperature employed in this synthesis so far), it was attempted to prepare a copper(I) compound, that would decompose to metallic copper at low temperatures. It is known, that cyclopentadienylcopper compounds[101] show interesting catalytic properties and serve as sources for copper of high purity, as they decompose at relatively low temperatures.[102] Thus, Saegusa, Ito et al. reported that the cyclopenta-dienylcopper(I) *tert*-butyl isocyanide complex catalyzes Michael-type additions of compounds containing active hydrogen, to acrylates and acrylonitrile.[103] Indeed, η^5-(cyclopentadienyl)trimethyl-phosphite-copper(I) at 20 °C efficiently catalyzes the reaction of **25**-Me with **168a** providing the pyrrole **173aa** in 89% yield within 2 h (Table 2, entry 14). However, all attempts to use this copper catalyst in the reaction of **25**-Et with both terminal and internal acetylenes without acceptor substituents, failed. Under the catalysis of CpCuP(OMe)$_3$ (5 mol%), the isocyanide **25**-Et dimerized to the imidazole **174**[39] in 85% yield at 20 °C within 16 h.

All catalysts (except nanosize-copper powder), which demonstrated moderate and good activity in the pyrrole formation (CuSPh, CuSePh, CuPPh$_2$, Cu$_2$O, Cu$_2$S), have one common feature, namely a σ-donating character of the counterion. Saegusa et al. reported that copper(I) *tert*-butoxide with similar electronic properties, reveals a strong affinity toward π-accepting ligands like isocyanides, and this has not been observed for common cuprous salts.[104] This feature of copper(I) compounds with σ-donating ligands may be ascribed to the enhancement of back-donation from the copper to π-accepting ligands, such as isocyanides, caused by increasing of electron density on Cu. Enhanced affinity of copper(I) compounds with σ-donating counterions to isocyanides appears to be crucial for the pyrrole formation. Cu(I)-Isocyanide complexes are known to abstract hydrogen from so-called active hydrogen compounds and to produce organocopper(I) isocyanide complexes,[37a, 105] which can undergo cycloadditions to form various heterocycles. In view of this, it is an open question, how copper(0) can be an active catalyst for the pyrrole formation. Metallic copper powder is known to dissolve in cyclohexyl isocyanide under an atmosphere of nitrogen to form a zero-valent copper-isocyanide complex, which can undergo an oxidative addition of a C-halogen bond.[106] Apart from the catalytic activity of metallic copper in the pyrrole synthesis demonstrated by de Meijere et al., Yamamoto et al. later reported, that metallic copper efficiently catalyzes the formation of imidazoles from two different isocyanides.[53] These results indicate, that copper(0) isocyanide, like copper(I) isocyanide complexes, are able to deprotonate compounds with active hydrogen.

To prove this hypothesis, the enantiomerically pure isocyanide **175**[107] was synthesized from *L*-isoleucine. Indeed, **175** underwent complete racemization upon heating at 85 °C for three hours in DMF in the presence of pre-activated copper nanoparticles (5 mol%).

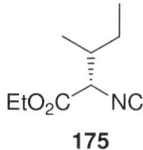

Figure 3. Chiral isocyanide **175**.

Kinetic Studies

Some simple kinetic studies were performed to determine the reaction order with respect to both the isocyanide **25**-Et and the acetylene **168a** in the Cu(I)-catalyzed pyrrole formation. The initial rates were estimated from the concentrations (determined from the ^{1}H NMR spectra employing hexamethylbenzene as an internal standard) of pyrrole formed

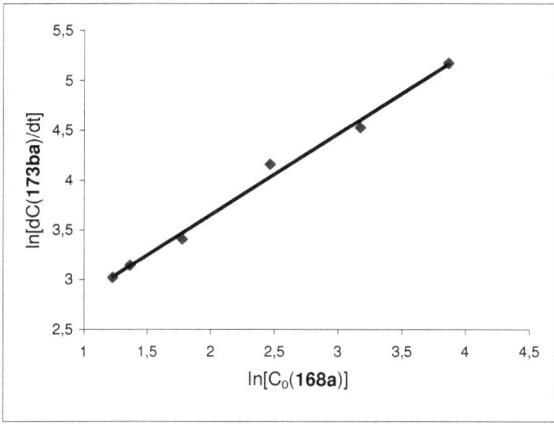

Figure 4. Determination of the reaction order with the respect to the acetylene **168a** in the initial phase of the formal cycloaddition of **25**-Et to **168a** in DMF at 85 °C catalyzed by CuSPh. C_0 (**168a**) = initial concentration of **168a**.

after 3 min each at constant initial concentrations of isocyanide **25**-Et (0.438 M) and different initial concentrations of acetylene **168a** (varying from 0.021 M to 0.291 M). The reactions were carried out in DMF at 85 °C. The dual logarithmic plot of ln[dC(**173ba**)/dt] versus ln[C_0(**168a**)] gave a straight line, the slope of which indicated (Figure 4) an order of 0.81 for this reaction with respect to the acetylene **168a**.

Analogously, the initial rates of the same reaction were estimated from the concentrations (determined from the ^{1}H NMR spectra employing hexamethylbenzene as an internal standard) of pyrrole formed after 3 min each at constant initial concentrations of **168a** (0.424 M) and different initial concentrations of **25**-Et (varying from 0.037 M to 0.183 M). The dual logarithmic plot (Figure 2) gave a straight line, the slope of which indicated an order of the reaction of 1.29 with the respect to the isocyanide **25**-Et.

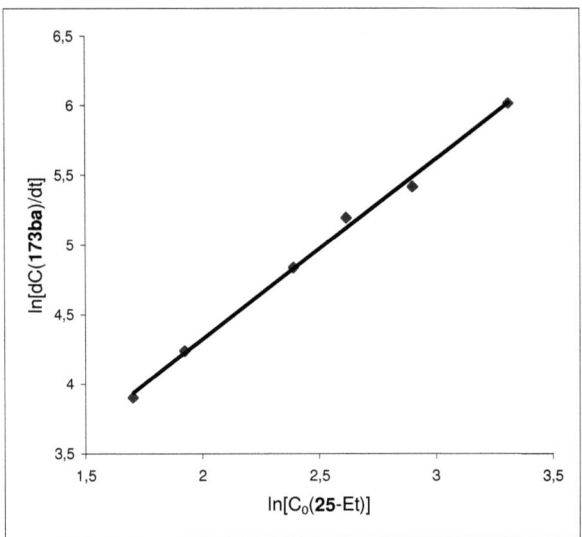

Figure 5. Determination of the reaction order with the respect to the isocyanide **25**-Et in the initial phase of the formal cycloaddition of **25**-Et to **168a** in DMF at 85 °C catalyzed by CuSPh. C_0 (**25**-Et) = initial concentration of **25**-Et.

These experimental data are in agreement with an overall second order of the reaction, i. e. first order with the respect to both, the acetylene and the isocyanide.

Synthesis of 2,3-Disubstituted Pyrroles

Although acetylenes without electron-withdrawing substituents have not been used earlier as cycloaddition-partners for isocyanides in pyrrole syntheses[50, 51] under usual conditions, an attempted reaction of 3-hexyne (**176**) with ethyl isocyanoacetate (**25**-Et) at elevated temperature (120 °C) in the presence of 1 equiv. of copper(I) iodide as a mediator and 5 equiv. of cesium carbonate as a base, gave a trace of the pyrrole **177** after 16 h.

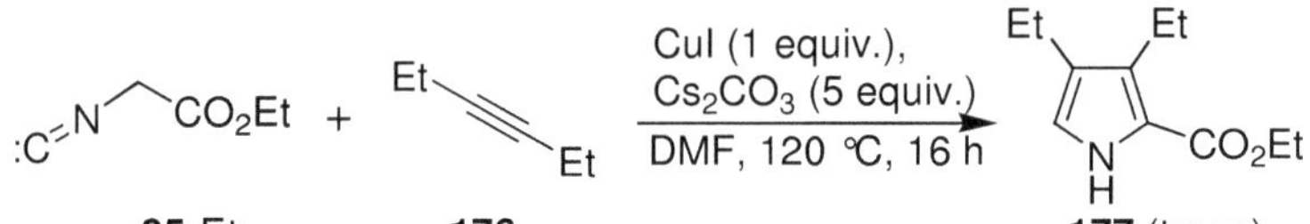

Scheme 45. Formation of pyrrole **177** from unactivated acetylene **176** and **25**-Et

Terminal acetylenes turned out to be more reactive under these conditions. Thus, ethyl 3-*n*-butylpyrrole-2-carboxylate **178ba** was obtained in 29% yield from 1-hexyne and ethyl isocyanoacetate (**25**-Et) (Table 3, entry 1). The best yield of **178ba** in this reaction was achieved at 120 °C, being almost the same as at 140 °C, while at 100 °C it was significantly lower (entries 3, 2, respectively). Different bases were tested, yet lithium and potassium carbonate were less effective than cesium carbonate, giving rise to 15 and 19% yield of **178ba** respectively, under the same conditions (entries 4, 5). Tertiary amines (Et$_3$N, EtNiPr$_2$, DBU, DABCO) were less effective than alkali carbonates, giving less than 10% yield of **178ba** under the same conditions. Although DMF was used as a solvent in most cases, *N*,*N*-dimethylacetamide worked as well (entry 6), in toluene **178ba** was obtained in a lower yield of 23% (entry 9). With catalytic quantities of CuI, only traces of **178ba** were isolated, while 1.3 equiv. of CuI did not provide an improvement compared to an equimolar quantity. Among the mediators used, CuOTf•0.5C$_6$H$_6$ was completely ineffective as well as Cu$_2$O, while CuI•P(OMe)$_3$ gave **178ba** in 10% yield. AgOAc was slightly worse (27% yield of **178ba**, entry 8) than CuI, and in view of the significantly lower prices of copper salts, no further silver mediators were tested. Surprisingly, copper(II) trifluoromethanesulfonate also achieved the formation of **178ba** in 21% yield (entry 7).

Table 3. Optimization of conditions for the synthesis of **178ba**.[a]

Entry	Mediator (equiv.)	Base (equiv.)	Solvent	Temperature [°C]	Yield[b] (%)
1	**CuI (1)**	**Cs_2CO_3 (5)**	**DMF**	**120**	**29**
2	CuI (1)	Cs_2CO_3 (5)	DMF	100	10
3	CuI (1)	Cs_2CO_3 (5)	DMF	140	28
4	CuI (1)	Li_2CO_3 (5)	DMF	120	15
5	CuI (1)	K_2CO_3 (5)	DMF	120	19
6	**CuI (1)**	**Cs_2CO_3 (5)**	**DMA**	**120**	**30**
7	Cu(OTf)$_2$ (1)	Cs_2CO_3 (5)	DMF	120	21
8	AgOAc (1)	Cs_2CO_3 (5)	DMF	120	27
9	CuI (1)	Cs_2CO_3 (5)	toluene	120	23

[a] All reactions were carried out with 1 mmol of the isocyanide **25-Et** and 5 mmol of the acetylene **167a** in 10 mL of solvent in a sealed vessel with stirring and heating for 12 h. [b] Yields of isolated product.

The yield of **178ba** could be further improved by gradually adding the isocyanide **25-Et** to a mixture of the copper mediator, cesium carbonate and the acetylene **167a** in DMF kept at 120 °C (Table 4). This procedure with a stoichiometric quantity of CuI provided the pyrrole **178ba** in 36% yield (entry 1). CuBr•SMe$_2$, CuBr and CuCl were equally effective, and all three of them were better than CuI (entries 2, 4, 5). But with a substoichiometric quantity (0.1 equiv.) of CuBr•SMe$_2$, only a trace of **178ba** was formed. The ratio of reagents had a big influence on the yield of pyrrole as well. The yields of **178ba** were best, when two and more equivalents of isocyanide were used, whereas with the ratio of 1.5 : 1 and 1 : 1 of **25-Et** to **167a**, the yields of **178ba** were 48 and 43%, respectively (entries 8, 9). Interestingly, with an excess of the acetylene **167a** (2 equiv.), **178ba** was obtained in 63% yield based on the isocyanide, indicating that the use of either an excess of the acetylene **167a** or an excess of the isocyanide **25-Et** are equally effective.

Table 4. Further optimization of conditions for the synthesis of **178ba**.[a],[b]

:C≡N−CH₂−CO₂Et + H−≡−Bu → pyrrole product

Scheme reactants: **25-Et** + **167a** → **178ba** (3-nBu, 2-CO₂Et pyrrole, NH)

Conditions: mediator, base, solvent

Entry	25-Et (equiv.)	167a (equiv.)	Mediator (equiv.)	Base (equiv.)	Yield[a] of 178ba (%)
1	1	1	CuI (1)	Cs$_2$CO$_3$ (5)	36[b]
2	5	1	CuBr·SMe$_2$ (1)	Cs$_2$CO$_3$ (5)	64[b]
3	1	1	CuBr·SMe$_2$ (0.1)	Cs$_2$CO$_3$ (5)	trace[b]
4	1	2	CuBr (1)	Cs$_2$CO$_3$ (3)	64[c]
5	1	2	CuCl (1)	Cs$_2$CO$_3$ (3)	64[c]
6	3	1	CuBr·SMe$_2$ (1)	Cs$_2$CO$_3$ (1)	70[b]
7	2	1	CuBr·SMe$_2$ (1)	Cs$_2$CO$_3$ (1)	70[b]
8	1.5	1	CuBr·SMe$_2$ (1)	Cs$_2$CO$_3$ (1)	48[b]
9	1	1	CuBr·SMe$_2$ (1)	Cs$_2$CO$_3$ (5)	43[b]
10	2	1	CuBr·SMe$_2$ (1)	Cs$_2$CO$_3$ (1)	63[b]
11	2	1	CuBr·SMe$_2$ (1)	Cs$_2$CO$_3$ (0.5)	trace[b]

[a] Yields of isolated product. [b] Method A: A solution of the isocyanide **25-Et** (1–5 mmol) in 5 mL of DMF was added dropwise at 120 °C within 2 h to a mixture of Cs$_2$CO$_3$, the copper acetylenide generated in situ from the acetylene **167a** and the copper(I) salt in 5 mL of DMF, and the mixture was stirred at 120 °C for 12 h. [c] Method B: A solution of the isocyanide **25-Et** (1 mmol) and the acetylene **167a** (1 mmol) in 5 mL of DMF was added dropwise within 2 h at 120 °C to a mixture of Cs$_2$CO$_3$, the copper acetylenide generated in situ from the acetylene **167a** (1 mmol) and the copper(I) salt in 5 mL of DMF, and the mixture was stirred at 120 °C for 12 h.

Table 5. Synthesis of 2,3-disubstituted pyrroles **178** and **179** from the isocyanide **25**-Et and terminal acetylenes **167**.[a],[b]

Entry	Acetylene	Product	Yield, (%)[a]
1	**167a** (H−≡−nBu)	**178ba** (3-nBu, 2-CO$_2$Et pyrrole)	70[b], 64[c]
2	**167b** (H−≡−CH$_2$OMe)	**178bb** (3-CH$_2$OMe, 2-CO$_2$Et pyrrole)	48[b], 45[c]
3	**167c** (H−≡−CH(OMe)−)	**178bc** (3-CH(OMe)−, 2-CO$_2$Et pyrrole)	74[b]
4	**167d** (H−≡−Ph)	**178bd** (3-Ph, 2-CO$_2$Et pyrrole)	40[b]
5	**167e** (H−≡−cyclopropyl)	**178be** (3-cyclopropyl, 2-CO$_2$Et pyrrole)	88[b]
6	**167f** (H−≡−tBu)	**178bf** (3-tBu, 2-CO$_2$Et pyrrole) and iso-**178bf** (4-tBu, 2-CO$_2$Et pyrrole)	5[c]

Table 5. (continued) Synthesis of 2,3-disubstituted pyrroles **178** and **179** from the isocyanide **25**-Et and terminal acetylenes **167**.[a],[b]

Entry	Acetylene	Product	Yield, (%)[a]
7	**167g** (H-C≡C-2-pyridyl)	**178bg** (3-(2-pyridyl)-pyrrole-2-CO₂Et)	16[b]
8	**167h** (H-C≡C-sec-Bu)	**178bh** (3-sec-butyl-pyrrole-2-CO₂Et)	58[b]
9	**167i** (H-C≡C-CH₂CH₂OH)	**179** (lactone-annelated pyrrole)	44[b], 37[c]

[a], [b], [c] See footnotes under Table 4

With the optimal conditions for the Cu(I)-mediated cycloaddition in hand, the reactions of ethyl isocyanoacetate (**25**-Et) with various terminal alkynes without acceptor substituents were carried out (Table 5). 1-Hexyne (**167a**) afforded the pyrrole **178ba** in 70 and 64% yield, respectively (entry 1), according to methods A and B (for details see footnotes under Table 4). 3-Methoxy-1-propyne (**167b**) with its donating methoxymethyl substituent, gave a lower yield of the pyrrole **178bb** (48%, entry 2). Bulky substituents R attached to the triple bond in **167** also led to decreased yields of the corresponding pyrroles **178**. Thus, **167h** with a *sec*-butyl group gave the pyrrole **178bh** in 58% yield (entry 8) compared to 70% of **178ba** (R = *n*-butyl). Phenylacetylene (**167d**), 2-pyridylacetylene (**167g**) and *tert*-butylacetylene (**167f**) afforded the corresponding pyrroles **178bd**, **178bg**, **178bf** / *iso*-**178bf** in 40, 16 and 5% yields, respectively (entries 4, 7, 6). In the latter case, a 5 : 1 mixture of the 2,3-**178bf** and the regioisomeric 2,4-disubstituted pyrrole *iso*-**178bf** was formed. The yields of pyrroles from cyclopropylacetylene (**167e**, entry 5) and from 3-methoxy-1-butyne (**167c**, entry 3) were the highest, although both of these acetylenes contain α-branched substituents. The cycloaddition of **25**-Et to 3-butyn-1-ol (**167i**) was accompanied by intramolecular transesterification of the ethoxycarbonyl group in the initial product, leading to the lactone-annelated pyrrole **179** in 44% yield (entry 9).

Various other acceptor-substituted isocyanides **63** were compared with **25**-Et in their CuBr-mediated formal cycloadditions to 1-hexyne (**167a**) (Table 6). With its bulky *tert*-butyl ester moiety, **25**-*t*Bu, gave a lower yield of **178ca** (47%, entry 2) than the ethyl ester **25**-Et gave **178ba** (70%, entry 1). *p*-Nitrophenylmethyl isocyanide (**63f**) afforded the corresponding pyrrole **178fa** in 20% yield only (entry 3). The methyl isocyanide with a diethylaminocarbonyl (**63g**), a dimethoxyphosphonyl (**63h**) and a *p*-toluenesulfonyl group (**41**-H) did not form any of the respective pyrroles at all, although the consumption of the isocyanide was complete in all these cases (entries 4–6). All 2,3-disubstituted pyrroles **178** obtained in this way were colorless solids or oils except for pyrrole **178fa**, which was isolated as red crystals. Indeed, a red color is typical for many other known 2-(4-nitrophenyl)-substituted pyrroles: 3,4-dimethyl-2-(4-nitrophenyl)-5-phenyl-1*H*-pyrrole and 2-(4-nitrophenyl)-3,4,5-triphenyl-1*H*-pyrrole,[108a] 2,5-bis-(4-nitrophenyl)-1*H*-pyrrole,[108b] 5-methyl-2-(4-nitrophenyl)-1*H*-pyrrole,[108c] 5-phenyl-2-(4-nitrophenyl)-1*H*-pyrrole.[108d]

Table 6. Synthesis of 2,3-disubstituted pyrroles **178** from various isocyanides **63** and 1-hexyne (**167a**).[a]

$$:C{\equiv}N{-}R \;+\; H{-}{\equiv}{-}Bu \xrightarrow[\text{DMF, 120 °C}]{\text{CuBr, Cs}_2\text{CO}_3} \text{pyrrole 178}$$

63 + 167a → 178 (3-nBu, 2-R pyrrole)

Entry	Isocyanide	Product	Yield,[b] (%)
1	CN–CH$_2$–CO$_2$Et **25-Et**	**178ba**	70
2	CN–CH$_2$–CO$_2$tBu **25-tBu**	**178ca**	47
3	CN–CH$_2$–C$_6$H$_4$–NO$_2$ **63f**	**178fa**	20
4	CN–CH$_2$–C(O)NEt$_2$ **63g**	**178ga**	0
5	CN–CH$_2$–P(O)(OMe)$_2$ **63h**	**178ha**	0
6	CN–SO$_2$Tol **41-H**	**178da**	0

[a] A solution of the isocyanide **63** (2 mmol) in 5 mL of DMF was added dropwise at 120 °C within 2 h to a mixture of Cs$_2$CO$_3$, the copper acetylenide generated in situ from the acetylene **167a** and CuBr in 5 mL of DMF, and the mixture was stirred at 120 °C for 1 h. [b] Yield of isolated product.

Mechanistic Considerations

A plausible mechanism of both the base-mediated and the copper(I)-catalyzed pyrrole formation from substituted methyl isocyanides **63** and electron-acceptor substituted alkynes **64** can be proposed (see Scheme 46).

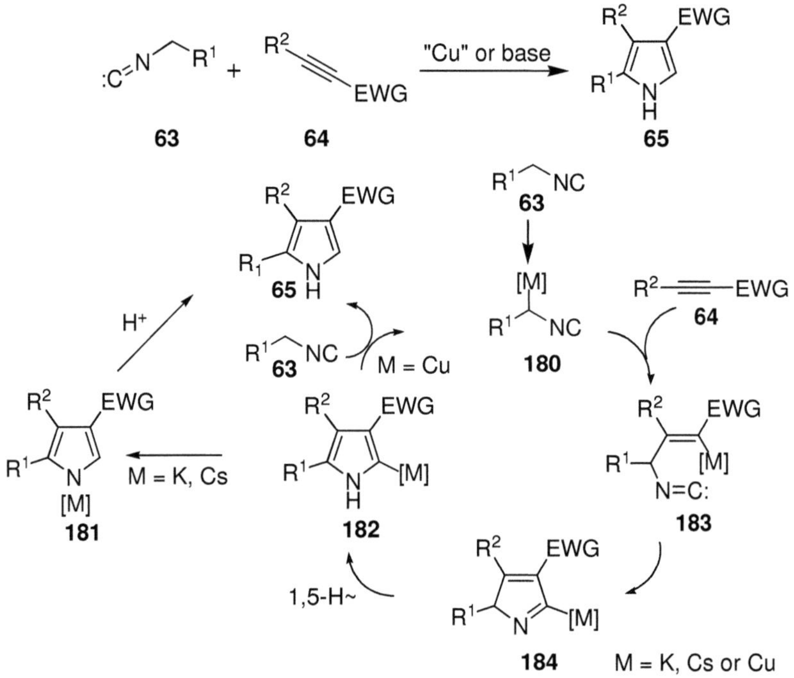

Scheme 46. Proposed mechanism for the formal cycloaddition of an α-metalated isocyanide **63** across the triple bond in an electron-deficient acetylene **64**.

The initiating step is the formation of an α-metalated isocyanide **180**. Not only Cu(I) compounds, but also metallic copper powder and Cu(II) salts (to some extent) are expected to be active in the formation of such a species. Subsequent Michael-type addition onto the triple bond of an activated acetylene **63** furnishes an unstable vinyl-organometallic compound **183**, which readily undergoes cyclization to the 2H-pyrroleninemetallic species **184**. The latter then experiences a 1,5-hydrogen shift to form **182**, and protonation of the latter by the isocyanide **63** gives the pyrrole **65**, completing the catalytic cycle for M = Cu. The intermediate **184** could also be protonated first, and then undergo a 1,5-hydrogen shift. There is no experimental evidence favoring either one of the two possibilities.

In the base-mediated pyrrole formation, Counterions like K^+ and Cs^+, which are harder than Cu^+ presumably lead to the *N*-metallated pyrrole **181**, which does not deprotonate to any significant extent a new molecule of isocyanide **63,** and this therefore requires a stoichiometric quantity of a base for the pyrrole formation in good yields.

The pyrrole formation in the copper(I)-mediated reaction between substituted methyl isocyanides **63** and unactivated terminal acetylenes **167** can be rationalized as follows (Scheme 47). Carbocupration[109] of the copper acetylenide **185** by the deprotonated isocyanide **180** followed by cyclization of the thus formed intermediate **186** would lead to the 2*H*-pyrrolenline-4,5-dicopper derivative **187**, which after 1,5-hydrogen shift and twofold protonation would give the pyrrole **178**.

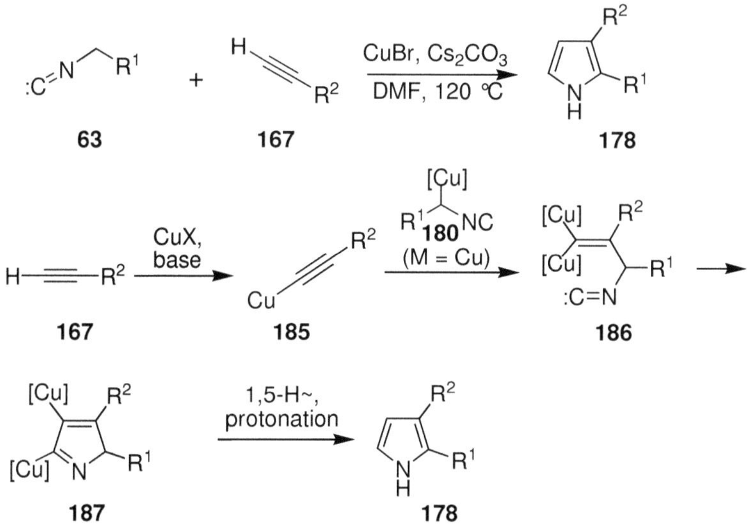

Scheme 47. Mechanistic rationalization of the copper(I)-mediated reaction of isocyanides **63** with a terminal acetylene **167** to yield a 2,3-disubstituted pyrrole **178**.

To support this hypothesis, hexynylcopper[110] (**185**, R^2 = *n*Bu) was prepared separately and treated with methyl isocyanoacetate (**25**-Me) in DMF at 120 °C, both in the presence of base and without it, to furnish the pyrrole **178aa** in 28 and 30 % yield, respectively. In addition, a reaction of **25**-Me with a twofold excess of 1-deutero-1-hexyne (**167a**-D) employing method B (see footnotes under Table 4) was carried out (Scheme 48).

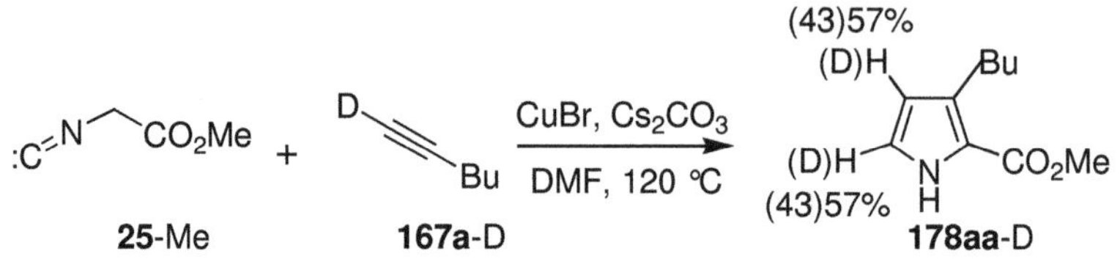

Scheme 48. Formation of the partly deuterated pyrrole **178aa**-D.

This reaction furnished a mixture of pyrroles **178aa**-D with approximately equal deuterium incorporation (43%) at positions 4 and 5, as evidenced by a ^{1}H NMR data. This fact confirms the intermediate formation of a 2H-pyrrolenline-4,5-dicopper species **187**, which is deuterated or protonated by **167a**-D or **25**-Me, respectively to give pyrrole **178aa**-D or **178aa**-H, respectively.

Conclusion

The direct formation of pyrroles from substituted methyl isocyanides **63** and acceptor- substituted acetylenes **64** under copper(I) catalysis represents a convenient route to 2,3,4-trisubstituted pyrroles and 3,4-disubstituted pyrroles **173** with sufficient functionality for further elaboration. The newly found route to 2,3-disubstituted pyrroles **178** from substituted methyl isocyanides **63** and non-activated terminal acetylenes (**167**) mediated by copper(I) compounds further enhances the versatility of these pyrrole syntheses. The prepared derivatives can also be easily transformed into pyrroles of higher or lower order of substitution according to established protocols.[111, 112]

2. ortho-Lithiophenyl Isocyanide: A Versatile Precursor to 3H-Quinazolin-4-ones and 3H-Quinazolin-4-thiones[113]

Background and Preliminary Considerations

Three major types of metallated isocyanides have been reported earlier (vide supra) to undergo subsequent cycloadditions to provide various heterocycles (Figure 6). Interestingly, ring metalated aryl(hetaryl) isocyanides have been elusive so far and hence the possibility of constructing heterocycles therefrom has not been explored. We envisaged that *ortho*-metallated phenyl isocyanides and their heteroanalogues could also be versatile precursors for certain types of heterocycles and considered that the elaboration of efficient method of its generating would be of great interest.

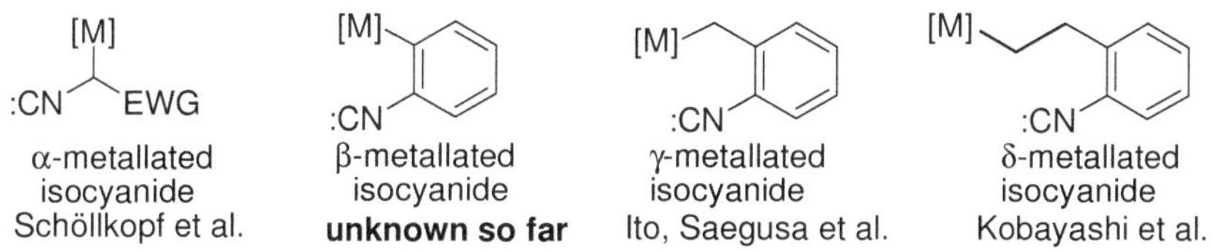

Figure 6. Different types of metallated isocyanides.

α-metallated isocyanide — Schöllkopf et al.
β-metallated isocyanide — **unknown so far**
γ-metallated isocyanide — Ito, Saegusa et al.
δ-metallated isocyanide — Kobayashi et al.

Thus, the reaction of *ortho*-metalated phenylisocyanide **188** with an isocyanate RNCO would provide *N*-metalated 2-isocyano benzamide **189** (X = O) capable of further cyclization to form metalated 3*H*-quinazolin-4-one **190**. The latter can be trapped in situ with various electrophiles to provide a convenient access to substituted 3*H*-quinazolin-4-ones **191** (Scheme 49).

This would be extremely useful, as 3*H*-quinazolin-4-ones have been reported to possess a vast range of biological activities, including analgesic, anti-Parkinsonian, CNS depressant, and CNS stimulating as well as tranquilizing, antidepressant, and anticonvulsant effects. Some of these compounds also act as psychotropic, hypnotic, cardiotonic, antihistamine agents,[114, 115] and possess cardiovascular activity as well as antiinflammatory activi-ty.[114, 116] Quinazolinones also inhibit monoamine oxidase, aldose reductase, tumor necrosis factor R, thymidylate synthase, pyruvic acid oxidation, as well

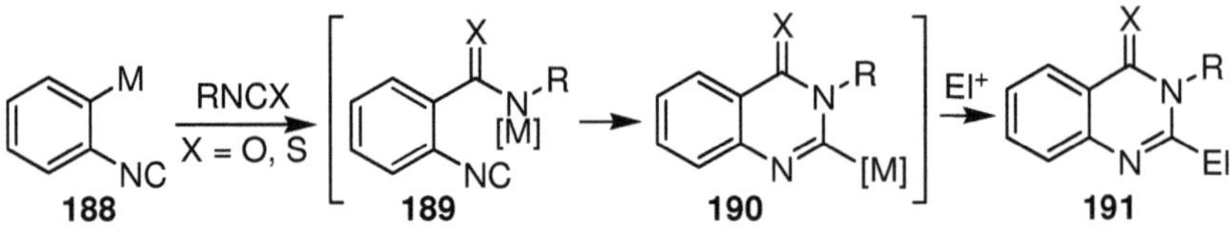

M
RNCX
X = O, S
NC
188
X
N
R
[M]
NC
189
X
N
R
N
[M]
190
El+
X
N
R
N
El
191

Scheme 49. A proposed approach to substituted 3*H*-Quinazolin-4-ones and 3*H*-Quinazolin-4-thiones **191** by a reaction of *ortho*-metallated phenyl isocyanide with isocyanates (isothiocyanates).

as acetylcholine-esterase activity and are antitumor, antiulcer, antiplatelet aggregation (glycoprotein IIb/ IIIa inhibitors),[117] and hypoglycemic agents.[114, 118] They are also potent antibacterial, antifungal, antiviral, antimyco-bacterial, and antimalarial agents.[114] Therefore, not surprisingly, they have been included in the list of molecules with "privileged structure"[119] for combinatorial chemistry, capable of binding to multiple receptors with high affinity.[120] Some derivatives of 3*H*-quinazolin-4-ones occur as natural products[121, 122] (Figure 7). Many of the numerous reported syntheses of these heterocycles start from anthranilic acid or its derivatives, but none of them uses the advantages of isocyanide chemistry.[123, 124]

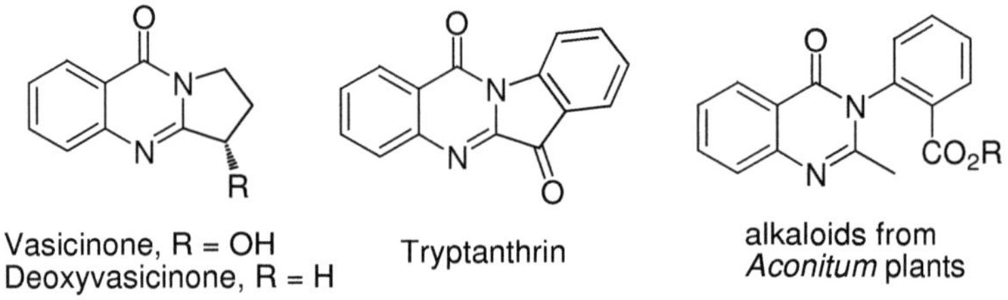

O
N
N
R
Vasicinone, R = OH
Deoxyvasicinone, R = H
O
N
N
O
Tryptanthrin
O
N
N
CO2R
alkaloids from
Aconitum plants

Figure 7. Some naturally occurring 3*H*-quinazolin-4-ones.[121,122]

Synthesis of 2-Substituted Phenyl Isocyanides by Reaction of *ortho*-Lithiophenyl Isocyanide with Electrophiles

To investigate the possibility of generating *ortho*-metallated phenyl isocyanide, two possible precursors for halogen–metal exchange reactions, *ortho*-bromo- and *ortho*-iodophenyl isocyanides **159**-Br and **159**-I were synthesized. The iodo derivative **159**-I turned out to undergo fast (<10 min) transmetalation reactions, when it was treated with *n*BuLi, *t*BuLi (–100 °C) or *i*PrMgCl•LiCl[125] (–78 °C) in THF. The target *ortho*-lithiophenyl isocyanide could also be obtained from the bromo derivative **159**-Br, synthesized from inexpensive 2-bromoaniline. The best and most reproducible results, in this case, were achieved with *n*BuLi in THF at –78 °C. Different electrophiles were tested in their reaction with *ortho*-lithiophenyl isocyanide (**188**-Li) generated *in situ* in this way (Table

7).[126] The respective 2-substituted phenyl isocyanides (**192**) were obtained in high yields (79–88%), except for 2-formylphenyl isocyanide **192c** (55%).

Table 7. Synthesis of 2-substituted phenyl isocyanides (**192**).

159-Br → 192, conditions: 1) nBuLi, 10 min; 2) E⁺, 3 h; −78 °C, THF

Electrophile	Product of type 192		Yield (%)[a]
I$_2$	(2-I-C$_6$H$_4$-NC)	**159-I**	88
ClCO$_2$Me	(2-CO$_2$Me-C$_6$H$_4$-NC)	**192a**	79
PhSSPh	(2-SPh-C$_6$H$_4$-NC)	**192b**	84
MeOCHO	(2-CHO-C$_6$H$_4$-NC)	**192c**	55
2-CO$_2$Me-C$_6$H$_4$-COCl	(2-NC-C$_6$H$_4$-CO-C$_6$H$_4$-2-CO$_2$Me)	**192d**	79

[a] Yield of isolated product

The standard reagent for the electrophilic installation of a formyl group, *N,N*-dimethyl formamide, in this case led to 2-(formylamino)-benzaldehyde **196**, which presumably was formed by base-catalyzed hydrolysis of the initially formed 1,3-benzooxazine derivative **194** under the aqueous work-up conditions (Scheme 50).

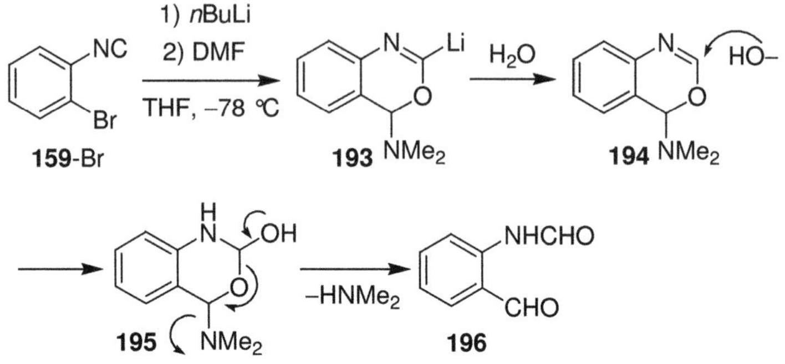

Scheme 50. Reaction of **159**-Br with *N,N*-dimethylformamide.

The 2-substituted phenyl isocyanides prepared in this way can be used for many purposes, particularly in multicomponent Ugi-Passerini reactions[6] or for the synthesis of correspondingly substituted anilines, to which isocyanides can easily be hydrolyzed under acidic conditions.[127]

Synthesis of Substituted 3*H*-Quinazolin-4-ones and 3*H*-Quinazolin-4-thiones

When isocyanates and isothiocyanates were employed as electrophiles, cyclic 3*H*-quinazolin-4-ones (-thiones) **191** were formed in high yields (69–91%) (Scheme 51).
Typically, the reactions with isocyanates were carried out at −78 °C and quenched with water at the same temperature, but in the case of isothiocyanates the mixtures were gradually warmed to −40 °C before quenching.
In contrast to these reactions of a *ortho*-lithiated phenylisocyanide, α-lithiated isocyanides have been reported mainly to give bis-adducts with isocyanates,[15b] indicating that the metalated five-membered heterocyclic intermediates formed in that case, were much more reactive than the lithiated derivatives of type **190**-Li formed from the β-lithiated isocyanide. This makes it possible to further diversify the 2-substituent of the 3*H*-quinazolin-4-ones (-thiones) **191** by trapping the intermediate **190**-Li with a second electrophile El^2X in situ.

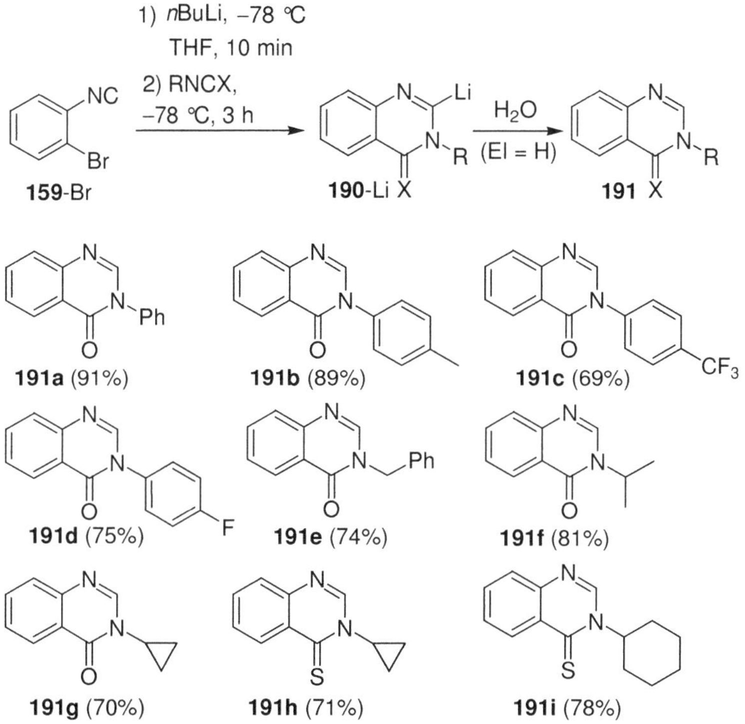

Scheme 51. Synthesis of 3*H*-quinazolin-4-ones 3*H*-quinazolin-4-thiones **191**.

Various 2,3-disubstituted 3*H*-quinazolin-4-ones **191j-o** could thus be conveniently prepared from 2-bromophenyl isocyanide **159**-Br in a three-step one-pot sequence (Table 8). 2-Halo-3*H*-quinazolin-4-ones of type **191m** have been reported to undergo substitution with nucleophiles[128] and also participate in different radical cyclization processes,[129] which opens an access to a large variety of substituted 3*H*-quinazolin-4-ones. Copper-catalyzed couplings of aryl thioethers of type **191k** with aryl iodides have also been reported.[130]

Table 8. The synthesis of 2,3-disubstituted 3H-quinazolin-4-ones **191**.

Reagents: 1) nBuLi, −78 °C, THF, 10 min; 2) R^{1}NCO, −78 °C, 3 h; 3) El^{2}X, −78 → 0 °C, 1 h. Starting material **159-Br** (2-bromobenzonitrile) → product **191**.

Isocyanate	Electrophile El^{2}X	Product of type **191**	Yield (%)[a]
PhNCO	ClCO$_2$Me	**191j** (El^2 = CO$_2$Me, R^1 = Ph)	73
PhNCO	PhSSPh	**191k** (El^2 = SPh, R^1 = Ph)	77
PhNCO	TosCN	**191l** (El^2 = CN, R^1 = Ph)	54
PhCH$_2$NCO	I$_2$	**191m** (El^2 = I, R^1 = CH$_2$Ph)	75
I(CH$_2$)$_3$NCO	—	**191n** (fused pyrrolo-quinazolinone)	72
2-(CO$_2$Me)C$_6$H$_4$NCO	—	**191o** (indolo-fused quinazolinone)	85

[a] Yield of isolated product.

Reactions of the lithiated intermediates of type **190**-Li with electrophiles can also occur intermolecularly, when the initially employed isocyanate already contains an appropriate functional group. Thus, 3-iodopropyl isocyanate and methyl 2-isocyanatobenzoate in one step gave 3*H*-quinazolin-4-ones **191n** and **191o** in 72 and 85% yield, respectively. Both deoxyvasicinone **191n**[131] and tryptanthrine **191o**[128a, 132] are naturally occurring alkaloids with important biological activities.

Conclusion

In conclusion, 2-substituted phenyl isocyanides are easily obtained by halogen–lithium exchange of *ortho*-bromophenyl isocyanide (**159**-Br) and subsequent trapping of the thus generated *ortho*-lithiophenyl isocyanide (**188**-Li) with electrophiles. This strategy has been effectively employed for the new three-step one-pot synthesis of substituted 3*H*-quinazolin-4-ones (-thiones) (**191**) including the naturally occurring alkaloids deoxyvasicinone (**191n**) and tryptanthrine (**191o**).

3. Reactions of ortho-Lithiophenyl (-Hetaryl) Isocyanides with Carbonyl Compounds – Rearrangements of 2-Metallated 4H-3,1-Benzoxazines[133]

Background and Preliminary Considerations

In the previous chapter, we reported that *ortho*-lithiophenyl isocyanide (**188**-Li), generated by bromine-lithium exchange on *o*-bromophenyl isocyanide (**159**-Br), can be employed for the synthesis of 2-substituted phenyl isocyanides **192** as well as 3*H*-quinazolin-4-ones and 3*H*-quinazolin-4-thiones **191** (Scheme 52).[113]

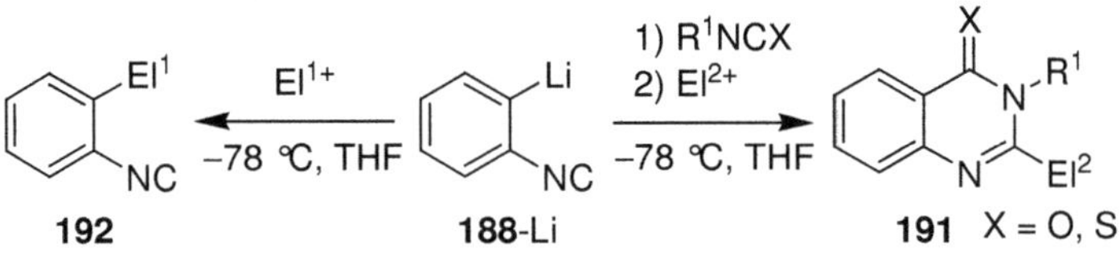

Scheme 52. Previously reported (Chapter 2) utilizations of *ortho*-lithiophenyl isocyanide (**188**-Li).[113]

For further elaboration of the chemistry of *ortho*-lithiophenyl isocyanide, we investigated its reactions with aldehydes, ketones, and carbon dioxide in details with the aim to develop an approach to substituted 4*H*-3,1-benzoxazines **201** and 4*H*-benzo[3,1]oxazin-4-ones **199**, respectively (Scheme 53). To broaden the scope of this method and to show its generality, we also intended to generate and employ in synthesis of various heterocycles some heteroanalogues **200** of *ortho*-lithiophenyl isocyanide **188**-Li.

Reactions of *ortho*-Lithiophenyl (-Hetaryl) Isocyanides with Carbonyl Compounds

Treatment of *ortho*-lithiophenyl isocyanide (**188**-Li) with aldehydes (**202a–i**) at −78 °C, and hydrolysis of the reaction mixture at the same temperature led to *ortho*-isocyanobenzyl alcohols **204** rather than the corresponding 4*H*-3,1-benzoxazines **201** (Table 9, entries 1–9). This may be due to a predominance of the initial alcoxide adduct **203** in the equilibrium with the lithiobenzoxazine **198**.

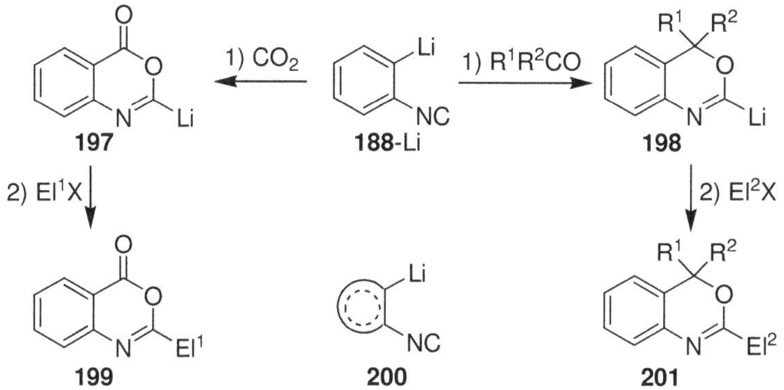

Scheme 53. Further elaboration of chemistry of *ortho*-lithiophenyl (-hetaryl) isocyanides.

The same behaviour was observed upon addition of α-metallated isocyanides and *ortho*-(lithiomethyl)phenyl isocyanides to carbonyl compounds and epoxides, which produced the respective acyclic isocyanoalcohols rather than corresponding 5-, 7- or 8-membered heterocycles **9**,[134] **94** and **96**,[60] respectively, (Figure 8) upon hydrolysis of the reaction mixture at low temperature.

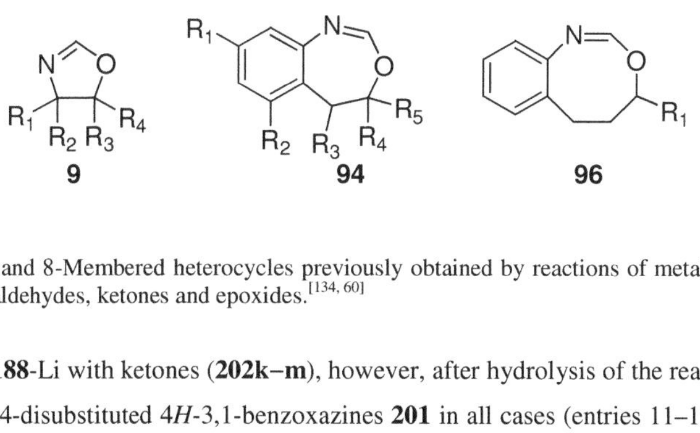

Figure 8. 5-, 7- and 8-Membered heterocycles previously obtained by reactions of metallated isocyanides with aldehydes, ketones and epoxides.[134, 60]

The reaction of **188**-Li with ketones (**202k–m**), however, after hydrolysis of the reaction mixture at –78 °C, led to 4,4-disubstituted 4*H*-3,1-benzoxazines **201** in all cases (entries 11–13). This may be caused by the Thorpe-Ingold conformational effect,[135] which places the alkoxide more closely to the isocyano group, and thus favors the cyclization of **203** to **198**.

Table 9. Reaction of **188**-Li with aldehydes and ketones succeeded by quenching with water at –78 °C.

Entry	Carbonyl Compound **202**	R^1	R^2	Product	Yield[a] (%)
1	a	Ph	H	**204a**	84
2	b	4-MeOC$_6$H$_4$	H	**204b**	83
3	c	4-ClC$_6$H$_4$	H	**204c**	89
4	d	4-pyridyl	H	**204d**	82
5	e	2-(5-methyl-thienyl)	H	**204e**	78
6	f	2-(5-methyl-furyl)	H	**204f**	88
7	g	tBu	H	**204g**	80
8	h	iPr	H	**204h**	36
9	i	1-(2-methyl-2-butene-1-yl)	H	**204i**	70
10	j	Me$_2$N	H	**196**	76
11	k	Ph	Ph	**201k**	48
12	l	Ph	CF$_3$	**201l**	78
13	m	Me	Me	**201m**	52

[a] Yields of isolated product.

It might also be due to enhanced thermodynamic stability of the cyclized **198** over the non-cyclized form **203** for the cases with $R^1, R^2 \neq H$. Upon treatment of **188**-Li with dimethylformamide and subsequent addition of water, 2-(formylamino)benzaldehyde (**196**) was isolated in 76% yield, apparently arising by hydrolysis of the initially formed 4*H*-3,1-benzoxazine **194**, as has previously been discussed (Chapter 2).[113] Obviously, in this case the NMe_2-donor group facilitated cyclization of **203** to **198**. Although 2-substituted 4*H*-3,1-benzoxazines are well-known compounds, simply accessible from the respective *o*-aminophenylcarbinols,[136] there is no generally applicable method for the synthesis of 2-unsubstituted heterocycles of type **201**.[137]

Yields of the final products **204** and **201**, respectively, were high from all non-enolizable aldehydes and ketones except for **201k** from benzophenone (**202k**), which has two large substituents, that might sterically encumber the addition of **188**-Li (entry 11). The yields of **204h** from isobutyraldehyde (**202h**) and of **204m** from acetone (**202m**) were significantly lower, probably because **188**-Li can abstract a proton from **202h** and **202m** in competition with adding to them (entries 8, 13). The reaction of *ortho*-lithiophenyl isocyanide **188**-Li with 3-methylbut-2-enal (**202i**) afforded the 1,2-adduct **204i** in 70% yield without traces of the 1,4-addition product (entry 9). Unsaturated alcohols of type **204i**, previously prepared by addition of substituted vinylmagnesium bromides to *N*-(*o*-acylphenyl)formamides, have been shown to undergo a Lewis acid-catalyzed cyclization followed by rearrangement to 1-formyl-1,2-dihydroquinoline derivatives.[138]

The adducts of *ortho*-lithiophenyl isocyanide (**188**-Li) to carbonyl compounds **202** can also be trapped with electrophiles other than water (Table 10). Thus, the initial adduct (**203d**) of **188**-Li to pyridine-4-carbaldehyde (**202d**) upon treatment with methyl chloroformate afforded the acyclic mixed methyl carbonate **205** in 56% yield (entry 1), while the adduct **198l** to 1,1,1-trifluoroacetophenone (**202l**) was trapped with methyl chloroformate and ethyl bromoacetate to furnish the 2-substituted 4*H*-3,1-benzoxazines **201l**-CO_2Me (45%) and **201l**-CH_2CO_2Et (47%), respectively (entries 2, 3). Addition of iodine to the same reaction mixture from **188**-Li and **202l** and subsequent aqueous work-up gave the oxo derivative **206** (77% yield, entry 4). The initially formed 2-iodo-4-phenyl-4-(trifluoromethyl)-4*H*-benzo[1,3]oxazine (**201l**-I) apparently undergoes rapid nucleophilic substitution by water and enol to ketone tautomerization during the work-up procedure and/or column chromatography on silica gel. The analogous 2-chloro derivative (**201l**-Cl), generated by treatment of the adduct of **188**-Li to **202l** with *N*-chlorosuccinimide as an electrophile, also could not be isolated and afforded **206**.

Table 10. Reaction of **188**-Li with carbonyl compounds **202** and trapping of the metallated intermediates with electrophiles other than water.

Entry	Carbonyl Compound **202** R^1 R^2			Electrophile ElX	Product	Yield[a] (%)
1	**d**	4-Py	H	MeO$_2$CCl	**205**	56
2	**l**	Ph	CF$_3$	MeO$_2$CCl	**201l-CO$_2$Me**	45
3	**l**	Ph	CF$_3$	EtO$_2$CCH$_2$Br	**201l-CH$_2$CO$_2$Et**	47
4	**l**	Ph	CF$_3$	I$_2$, then H$_2$O[b]	**206** (from **201l-I**)	77
5	**l**	Ph	CF$_3$	I$_2$, then morpholine[c]	**207** (from **201l-I**)	55

[a] Yield of isolated product. [b] Aqueous work-up procedure. [c] Morpholine was added before aqueous work-up.

The same reaction mixture from **188**-Li, **202l** and iodine upon treatment with morpholine prior to the aqueous work-up led to 2-morpholinylbenzoxazine **207** in 55% yield (entry 5).
ortho-Lithiophenyl isocyanide **188**-Li also reacts with carbon dioxide at −78 °C to initially form lithium *ortho*-isocyanobenzoate **208**, which equilibrates with the 2-lithiobenzoxazin-4-one **197**. The latter reacts with iodine to furnish 2-iodobenzoxazin-4-one (**199**-I) which readily undergoes in situ substitution by added nucleophiles such as water, morpholine and aziridine to provide the correspondingly substituted 4-*H*-benzo[3,1]oxazin-4-ones[139, 140] **199**-Nu and isatoic anhydride **209** in a one-pot four-step procedure in moderate yields (Scheme 54).

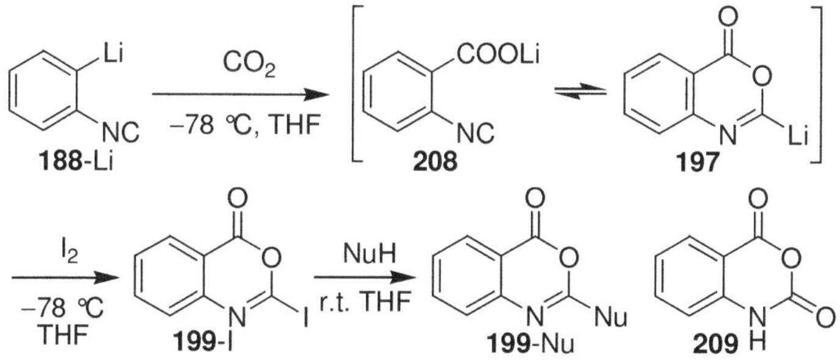

NuH = morpholine; **199**-morph (45%)
NuH = aziridine; **199**-azirid (50%)
NuH = H_2O; **209** (61%)

Scheme 54. Synthesis of 2-substituted 4*H*-benzo[d][1,3]oxazin-4-ones (**199**-Nu) and isatoic anhydride **209**.

Copper(I)-catalyzed Cyclizations of Isocyanobenzyl alcohols 204

The isocyanobenzylalcohols **204** with R^2 = H obtained from **188**-Li and aldehydes were found to undergo cyclization to the corresponding 4*H*-3,1-benzoxazines **201** under Cu_2O catalysis in high yields (Table 11, entries 1−5) in the same way, as it had previously been demonstrated for the synthesis of 4,5-dihydro-3,1-benzoxazepines **94** and 4*H*-5,6-dihydro-3,1-benzoxazocines **96**.[60]

Table 11. Cu₂O-Catalyzed cyclization of isocyanobenzylalcohols **204**.

Entry	Isocyano-alcohol 204	Product	Yield[a] (%)
1	204a	201a	86
2	204b	201b	74
3	204c	201c	75
4	204d	201d	73

Table 11. (continued) Cu$_2$O-Catalyzed cyclization of isocyanobenzylalcohols **204**.

Entry	Isocyano-alcohol **204**	Product	Yield (%)
5	**204g**	**201g**	83
6	**204f**	**210f**	66
7	**204h**	**210h**	68
8	**212d**	**211d**	74[b]

[a] Yield of isolated product. [b] Total yield for addition and subsequent cyclization, the crude isocyanoalcohol **212d** was used for the transformation without purification.

Treatment of the isocyanobenzylalcohols **204** with bases such as DBU and KO*t*Bu also led to the target 4*H*-3,1-benzoxazines **201**, although in lower yields. Such 2-unsubstituted compounds turned out to be unstable in acidic as well as in basic media, but could be isolated by flash chromatography on silica gel pretreated with triethylamine.

In the cases of the isocyanobenzylalcohols **204f**, **204h** and **212d** (the latter was used in the cyclization step directly after its formation from 3-isocyano-2-lithiothiophene (**216**) and pyridine-4-carbaldehyde (**202d**) without purification by column chromatography) the arene-annelated tetrahydrofuranimines **210f,h** and **211d**, respectively, were obtained unexpectedly as the sole products. Products of this type and indolin-2-ones **215** were also formed upon warming to ambient temperature of the reaction mixtures after the addition of *ortho*-lithiophenyl isocyanide (**188**-Li) and *ortho*-lithiohetaryl isocyanides **216** and **218** to various carbonyl compounds (Table 12). The latter two organolithium reagents were generated with equal ease as **188**-Li from the corresponding bromohetaryl isocyanides.

Novel Rearrangements of 2-Metallated 4*H*-3,1-Benzoxazines

Apparently, the lithiated intermediates of type **198** can undergo ring contraction to form the lithiated precursors of **213** or **214** upon warming to ambient temperature of the reaction mixture obtained after addition of lithiated isocyanides **200** to carbonyl compounds. All three compounds of type **198** with trifluoromethyl substituents obviously rearranged to iminophthalanes **210o** and its heteroanalogues **217l**, **219l**, respectively (Table 12). The other examples only furnished indolin-2-ones **215n**, **215k** and **217k**, respectively. Compound **215n** was isolated after the reaction of **188**-Li with pyridine-2-carbaldehyde (**202n**) and subsequent treatment of the reaction mixture with water at −78 °C. In this case, the coordination of lithium by the pyridyl nitrogen may have played a crucial role in shifting of the equilibrium from **203** to **198** and facilitate the rearrangement to the lithiated precursor of **214**. Indolin-2-ones (2-oxoindoles) of type **215** represent an important class of heterocycles with a wide range of biological activities,[141] while only a few isobenzofuran-1(3*H*)-imines (iminophthalanes) of type **200** have been described previously.[142]

Table 12. Addition of *ortho*-lithioaryl isocyanides to aldehydes and ketones with subsequent rearrangement.

Reaction scheme: *o*-lithiated aryl isocyanide **200** + $R^1R^2C=O$ → 1) −78 °C, THF; 2) −78 °C to r.t; 3) H_2O → product **213** (isobenzofuranone-type, NH) or **214** (indolin-2-one-type, NH).

o-Lithiated Aryl Isocyanide	Carbonyl Compound	R^1	R^2	Product	Yield[a] (%)
188-Li	**202n**	2-pyridyl	H	**215n** (3-(2-pyridyl)indolin-2-one)	79[b]
188-Li	**202o**	Me	CF_3	**210o** (3-Me-3-CF_3 isobenzofuran-1(3H)-imine, NH)	58
188-Li	**202k**	Ph	Ph	**215k** (3,3-diphenylindolin-2-one)	42
216 (2-Li-3-NC-thiophene)	**202k**	Ph	Ph	**217k** (thieno-fused, 3,3-Ph,Ph, C=O, NH)	52
216	**202l**	Ph	CF_3	**211l** (thieno-fused, 3-Ph-3-CF_3, NH)	75
218 (2-Li-3-NC-pyridine)	**202l**	Ph	CF_3	**219l** (pyrido-fused, 3-Ph-3-CF_3, NH)	64

[a] Yield of isolated product. [b] The reaction mixture was treated with H_2O at −78 °C.

Mechanistic Considerations

The only known ring contraction of 2,4-diarylsubstituted 4*H*-3,1-benzoxazines with formation of 3*H*-indol-3-ols proceeds in strongly basic media and was rationalized mechanistically as an intramolecular nucleophilic addition of 4-deprotonated benzoxazine **220** followed by epoxide opening (Scheme 55).[143] However, the above mentioned benzotetrahydrofuranimines and indolin-2-ones obviously cannot be formed in such a way.

Formally, the isobenzofuranimine of type **210** and the indolin-2-one of type **215**, respectively, could arise from the initially formed 2-lithium 4*H*-3,1-benzoxazine of type **198** by a [1,2]-migration of the aryl group next to nitrogen or of the alkyl group next to the oxygen atom and subsequent protonation.

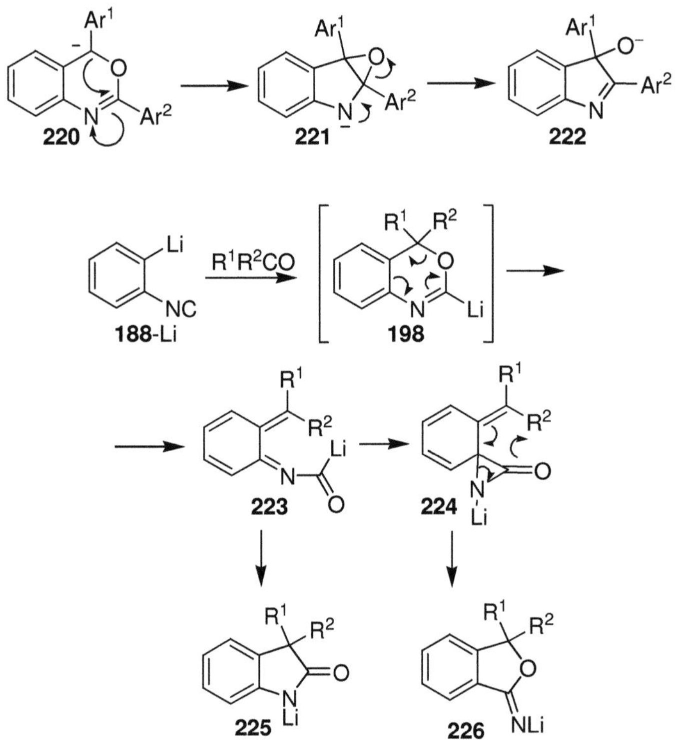

Scheme 55. A known ring contraction of 2,4-diaryl-3,1-benzoxazines[143] and proposed mechanism for the reaction of **188**-Li with ketones.

Rearrangements with [1,2]-migration of an alkyl group from an oxygen and from a quaternary nitrogen atom to an adjacent carbanion center have been known for quite some time as Wittig[144] and Stevens rearrangements,[145] respectively. Yet, the stereoelectronic requirements make such [1,2]-migrations unlikely in the case of **198**. More probably, the intermediate **198** undergoes a pericyclic ring opening to yield **223**, which by intramolecular 1,4-addition would furnish the lithiated indolin-2-one **225** or by intramolecular 1,2-addition and subsequent 6π-pericyclic reaction of the resulting **224** provide the lithiated isobenzofuranimines **226** (Scheme 55).[146]

On the other hand, in the Cu$_2$O-catalyzed transformations of isocyanobenzylalcohols **204** to **201** and **210**, the process starts with the coordination of an isocyano group to the Cu(I) species, and this is succeeded by nucleophilic addition of the hydroxyl group to thus activated isocyano group in **227** to yield, after deprotonation, the metallated 4H-3,1-benzoxazine **228** (Scheme 56).

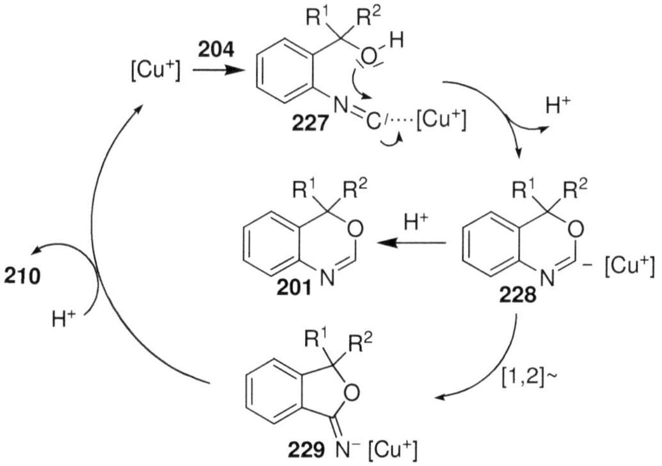

Scheme 56. Mechanism of the Cu$_2$O-catalyzed cyclization of isocyanobenzylalcohols **204**.

The latter rearranges just like **198** to provide the deprotonated isobenzofuran-1(3H)-imine **229** which, after protonation, forms **210**. Alternatively, the intermediate **228** can be protonated directly to yield the 4H-3,1-benzoxazine **201** as was also observed experimentally. The transformation of **227** to **228** may be also regarded as an isocyanide insertion into an O–Cu bond, but not as its insertion into O-H linkage[147] because the product of such a process, the 4H-3,1-benzoxazine **201** is not converted to **210** under the same reaction conditions, as was confirmed by a control experiment.[148] Interestingly, the predominant formation of **201** or **210** from **204** is intricately controlled by the type of substitution. Thus, **204h** (R^1 = iPr, R^2 = H) gave the isobenzofuran-1-(3H)-

imine **210h**, while **204g** (R^1 = *t*Bu, R^2 = H) provided the corresponding benzoxazine **201g** exclusively. The isocyanobenzylalcohols **204f** and **212d** with furyl and thienyl moieties, afforded selectively isobenzofuran-1-(3*H*)-imine **210f** and thiophene-annelated tetrahydrofuranimine **211d**, respectively, whereas all other aryl-substituted isocyanobenzylalcohols **204a–204d** gave the corresponding 4*H*-3,1-benzoxazines **201**.

Conclusion

In conclusion, the reactions of *ortho*-lithiophenyl isocyanide (**188**-Li) and other *ortho*-lithiohetaryl isocyanides (**216**, **218**) with aldehydes, ketones and carbon dioxide furnish, apart from the expected isocyanobenzylalcohols **204**, 4*H*-3,1-benzoxazines **201** and 4*H*-benzoxazin-4-ones **199**-Nu, also iminophthalanes of type **210** or indolin-2-ones of type **215**, respectively, by two novel rearrangements of the intermediate 2-lithio 4*H*-3,1-benzoxazines (**198**).

4. Synthesis of 1-Substituted Benzimidazoles from o-Bromophenyl Isocyanide and Amines[149]

Background and Preliminary Considerations

Compounds containing a benzimidazole moiety possess a wide range of biological activities and therefore represent "privileged" structures having a significant importance in medicinal chemistry.[120, 150] In fact, certain compounds of this type with high activity against Hepatitis B and C viruses have been identified,[151] others have been found to be potent lymphocyte specific kinase (Lck) inhibitors,[152] nonpeptide thrombin inhibitors,[153] and antiallergic agents.[154] The classical construction of the five-membered heterocycle in benzimidazoles involves the reaction of an o-phenylenediamine with a carboxylic acid or one of its equivalents under harsh dehydrating conditions.[155] Alternatively, several transition metal-catalyzed syntheses of benzimidazoles and related systems have been reported recently.[156] The following precursors have been typically used so far: 2-haloacetanilides, N-substituted amidines, N-substituted N'-(2-halophenyl)amidines and -guanidines, N-substituted N'-(2-halophenyl)ureas and -thioureas to give the corresponding 2-substituted benzimidazoles. We envisaged, that a different convenient access to *2-unsubstituted* benzimidazoles, which remained elusive so far, could start from primary amines and *ortho*-haloaryl isocyanides, which have already shown their versatility as building blocks for various other heterocycles.[88, 113, 133] Isocyanides are known to react with amines in the presence of copper[157] as well as other metal salts[158] to form amidines in excellent yields. Amidines formed from *ortho*-haloaryl isocyanides in such a way ought to be able to undergo an intramolecular copper-catalyzed N-arylation[159] to furnish benzimidazoles. As both steps require the same type of catalyst, one ought to be able to perform them sequentially in a one-pot operation. This process would provide synthetically useful 2-unsubstituted benzimidazoles which can be further elaborated by attaching various substituents at the 2-position e. g. by means of lithiation/electrophilic substitution[160] or transition metal-catalyzed C-H-activation[161] as well as cross-coupling reactions[162] of the easily accessible corresponding 2-halobenzimidazoles.[163]

Optimization of the Reaction Conditions for the Synthesis of 1-Benzylbenzimidazole

The reaction of *o*-bromophenyl isocyanide (**159**-Br) and benzylamine (**230a**) was chosen as a model system for the optimization of reaction conditions (Table 13). Cesium carbonate was found to be the best base, giving higher yields of 1-benzylbenzimidazole (**232a**) than potassium carbonate (entry 3), potassium phosphate (entry 1), lithium or potassium *tert*-butoxides (entries 4 and 5, respectively) and triethylamine (entry 6).

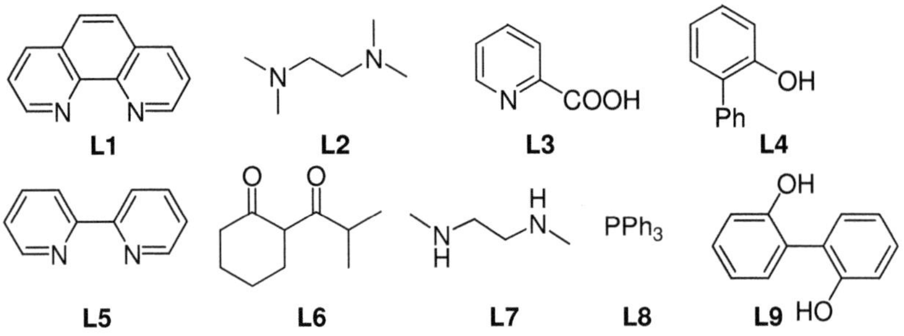

Figure 9.   Ligands tested for the synthesis of **232a** (see Table 13).

With the latter, the formamidine **231a** (52%) was isolated as the main product along with the benzimidazole (**232a**) in low yield (11%). Formation of **231a** was also observed in other cases, in which **232a** was obtained in low yields. This indicates that the initially proposed sequence of an α-addition of the amine to the isocyanide and subsequent intramolecular amination is operational, and apparently the second step is more affected by the conditions used. Dimethylformamide turned out to be the solvent of choice, as the reaction in other solvents (DME, dioxane, toluene, DMSO) afforded **232a** in lower yields (entries 7–10). Various ligands **L1–L9** (Figure 9) usually employed in copper-catalyzed arylations of amines, have been tested. 1,10-Phenanthroline (**L1**) and 2-phenylphenol (**L4**) furnished the best results (68 and 65% yield of **232a**, entries 2 and 14, respectively), although the ligand effect was not as significant as one would have imagined. Replacement of CuI by CuBr has almost not changed the yield of **217a** (68 versus 70%, entries 2 and 20), while Cu$_2$O was far less effective (entry 21).

Table 13. Optimization of the reaction conditions for the synthesis of 1-benzylbenzimidazole (**232a**).

Entry	Catalyst	Ligand	Base	Solvent	Yield (%)[a]
1	CuI	L1	K$_3$PO$_4$	DMF	56
2	CuI	L1	Cs$_2$CO$_3$	DMF	68
3	CuI	L1	K$_2$CO$_3$	DMF	27
4	CuI	L1	LiOtBu	DMF	58
5	CuI	L1	KOtBu	DMF	36
6	CuI	L1	Et$_3$N	DMF	11[b]
7	CuI	L1	Cs$_2$CO$_3$	DME	54
8	CuI	L1	Cs$_2$CO$_3$	dioxane	37
9	CuI	L1	Cs$_2$CO$_3$	toluene	14
10	CuI	L1	Cs$_2$CO$_3$	DMSO	42
11	CuI	L2	Cs$_2$CO$_3$	DMF	56
12	CuI	L3	Cs$_2$CO$_3$	DMF	49
13	CuI	none	Cs$_2$CO$_3$	DMF	38
14	CuI	L4	Cs$_2$CO$_3$	DMF	65
15	CuI	L5	Cs$_2$CO$_3$	DMF	58
16	CuI	L6	Cs$_2$CO$_3$	DMF	59
17	CuI	L7	Cs$_2$CO$_3$	DMF	40
18	CuI	L8	Cs$_2$CO$_3$	DMF	57
19	CuI	L9	Cs$_2$CO$_3$	DMF	32
20	CuBr	L1	Cs$_2$CO$_3$	DMF	70
21	Cu$_2$O	L1	Cs$_2$CO$_3$	DMF	26

[a] Yield of isolated product **232a**. [b] In addition, the intermediate **231a** was also isolated in 58% yield.

Different methods of reagents addition to the reaction mixture as well as temperature conditions have been tested. Thus, when a solution of benzylamine (**230a**) and **159**-Br was slowly added at

90 °C to the mixture of the other reagents, no benzimidazole (**232a**) was formed at all. Carrying out the operation first at r.t. within 2 h, then gradually (within 30 min) warming the mixture to 90 °C, and then keeping it at the same temperature for 14 h, gave the best yields of **232a**. Other *o*-halophenyl isocyanides were also tested towards the same transformation. *o*-Chlorophenyl isocyanide (**159**-Cl) did not provide the corresponding benzimidazole neither under the best conditions found for **159**-Br nor at increased temperatures up to 140 °C. *o*-Iodophenyl isocyanide (**159**-I) with benzylamine (**230a**), on the contrary, furnished benzimidazole **232a** even at 50 °C, but in 22% yield only. The conditions optimized for **159**-Br (90 °C) applied to **159**-I, gave **232a** in 40% yield. Accordingly, it was not considered meaningful to test other temperatures for **159**-I, and work was focused on the use of **159**-Br for the synthesis of benzimidazoles **232**.

Scope and Limitations of the Synthesis

Employing the optimized conditions for **232a**, various *N*-substituted benzimidazoles **232b-l** have been synthesized from *o*-bromophenyl isocyanide (**159**-Br) and primary amines **230b-l** (Table 14). *n*-Alkylamines and benzylamines in general gave slightly better yields of benzimidazoles **232** than *sec*-alkylamines like cyclopropylamine and cyclohexylamine (entries 10 and 11, respectively), while 2-methoxybenzylamine with an *ortho*-substituent still afforded the corresponding benzimidazole **232f** in 65% yield (entry 6). Amines with decreased nucleophilicity, such as 4-trifluoromethylbenzylamine and 4-methylaniline (entries 9 and 12), furnished the corresponding benzimidazoles in slightly lower yields (55, 40%, respectively). The twofold reaction of ethylenediamine with **159**-Br afforded the 1,2-di(benzimidazolyl)ethane **232e** in 42% yield (entry 5). The reaction of *o*-bromophenyl isocyanide (**159**-Br) with *tert*-butylamine **230m** surprisingly did not provide *N-tert*-butyl benzimidazole **232m** at all. The major product, isolated in 38% yield, was identified as 1-(2-bromophenyl)benzimidazole (**232n**).

Table 14. The synthesis of N-substituted benzimidazoles **232**.

Entry	RNH₂	Product	Yield, (%)[a]
1	BnNH₂ (**230a**)	**232a** (R = Ph)	70
2	nPrNH₂ (**230b**)	**232b** (R = nPr)	65
3	BnO(CH₂)₃NH₂ (**230c**)	**232c** (R = (CH₂)₃OBn)	66
4	3-(2-aminoethyl)-N-methylindole (**230d**)	**232d**	59
5	H₂N(CH₂)₂NH₂ (**230e**)	**232e** (bis-benzimidazole)	42
6	2-MeO-C₆H₄CH₂NH₂ (**230f**)	**232f**	67

Table 14 (continued). The synthesis of *N*-substituted benzimidazoles **232**.

Entry	RNH₂	Product	Yield, (%)[a]
7	MeO-C₆H₃(OMe)-CH₂NH₂ **230g**	**232g**	65
8	furfurylamine **230h**	**232h**	46
9	4-F₃C-C₆H₄-CH₂NH₂ **230i**	**232i**	55
10	cyclopropyl-NH₂ **230j**	**232j**	40
11	cyclohexyl-NH₂ **230k**	**232k** cHex	46
12	4-Me-C₆H₄-NH₂ **230l**	**232l** pTol	41
13	tBu-NH₂ **230m**	**232n** (with Br)	38[b]

[a] Yield of isolated product. [b] Only the depicted product **232n** was isolated and identified

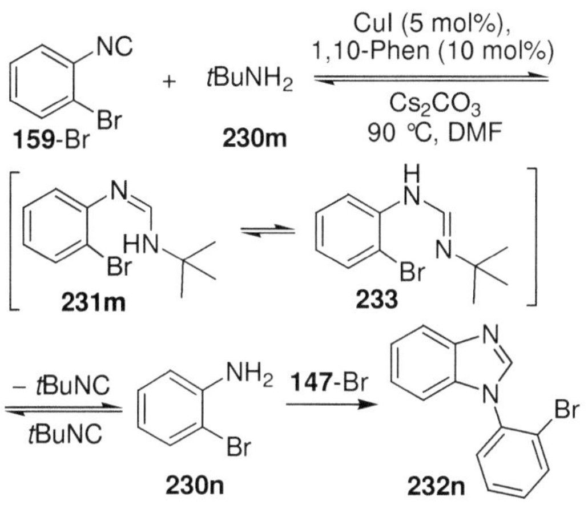

Scheme 57. Proposed mechanism for the formation of the benzimidazole **232n**.

The formation of **232n** (Scheme 57) can be rationalized assuming a reversible addition of *tert*-butylamine onto the isocyano group of **159**-Br. Similar reversible additions of *N*-unsubstituted indoles onto aryl isocyanides have previously been observed in a ruthenium-catalyzed formation of indoles.[64] The corresponding formamidine **231m**, due to its bulky *tert*-butyl group does not undergo cyclization to the *N*-*tert*-butylbenzimidazole (**232m**), but equilibrates under the basic reaction conditions with its tautomer, the formamidine **233**, which would reversibly release *tert*-butyl isocyanide and form *o*-bromoaniline (**230n**). The latter would react with *o*-bromophenyl isocyanide **159**-Br, still existing in the reaction mixture, just as 4-methylaniline does (see Table 14, entry 12), irreversibly forming benzimidazole **232n**. In a control experiment, the reaction of **159**-Br with *o*-bromoaniline (**230n**) under the same conditions also provided the benzimidazole **232n** in 42% yield.[164]

To broaden the scope of the new method, 2-bromo-3-isocyanothiophene (**234**) was employed in the copper-catalyzed reaction with amines. Indeed, the three examples **235a**, **235c** and **235d** of the the less common 3-substituted 3*H*-thieno[2,3-d]imidazoles **235** (Scheme 58) were isolated albeit in slightly lower yields (49, 44 and 44%, respectively) than the corresponding benzimidazoles.

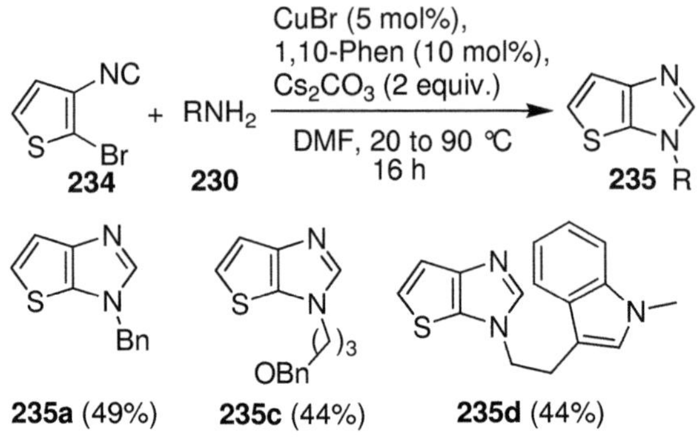

Scheme 58. The synthesis of 3-substituted 3*H*-thieno[2,3-d]imidazoles **235**.

Conclusion

In conclusion, a novel copper-catalyzed synthesis of benzimidazoles from *o*-bromoaryl isocyanides and primary amines has been developed. This new sequential reaction consisting of a copper-catalyzed addition of an amine onto an isocyano group followed by a copper-catalyzed intramolecular arylation of a thus formed amidine provides a convenient access to 1-substituted benzimidazoles **232** and related 3-substituted 3*H*-thieno[2,3-d]imidazoles **235** in moderate to good yields

C. Experimental Section

General

Reagents and Chemicals

Diethyl ether, tetrahydrofuran, 1,2-dimethoxyethane, benzene, and toluene were distilled from sodium benzophenone ketyl, dichloromethane and dimethylformamide from molecular sieves 4 Å, acetonitrile from P_4O_{10}.

Commercial nanosize copper powder (Aldrich) was preactivated by heating *in vacuo* (0.05 mbar) at 150 °C overnight, and was stored under Ar. The activity of the thus prepared catalyst does not deteriorate within at least 2 weeks.

The following compounds were prepared according to the corresponding literature procedures:
Methyl isocyanoacetate (**25**-Me),[165] ethyl isocyanoacetate (**25**-Et),[165] *tert*-butyl iso-cyanoacetate (**25**-*t*Bu),[165] methyl cyclopropylpropiolate (**168a**),[166] 4-nitrophenyl-methyl isocyanide (**63f**),[167] methyl 3-(4-fluorophenyl)propiolate (**168d**),[168] methyl 3-(4-trifluo-romethylphenyl)-propiolate (**168e**),[169] methyl 3-(thiophen-2-yl)propiolate (**168g**),[170] methyl 3-(pyridin-2-yl)propiolate (**168f**),[171] cyclopropylacetylene (**167e**),[172] *N,N*-diethyl-2-isocyanoacetamide (**63g**),[173] dimethyl 2-isocyano-1-oxoethylphospho-nate,[174] CpCuP(OMe)$_3$,[101] 1-deutero hexyne-1 (**167a-D**),[175] cyclopropylisocyanate,[176] 3-iodo-propyl isocyanates,[177] 3-aminothiophene,[178] 3-(benzyloxy)propyl-1-amine (**230c**).[179]

All other chemicals were used as commercially available.

Separation and Identification of the Compounds

Chromatography: Analytical TLC was performed on 0.25 mm silica gel 60F plates (Macherey-Nagel) with 254 nm fluorescent indicator from Merck. Plates were visualized under ultraviolet light and developed by treatment with the molybdenephosphoric acid solution. Chromatographic purification of products was accomplished by flash column chromatography, as described by Still and coworkers[180] on Merck silica gel, grade 60 (0.063–0.200 mm, 70–230 mesh ASTM)

NMR: Nuclear magnetic resonance (^{1}H and ^{13}C NMR) spectra were recorded at 250, 300, or 500 (^{1}H), 62.9, 75.5, or 125 [^{13}C, APT (Attached Proton Test)] MHz on Brucker AM 250, Varian Unity-300, AMX 300 and Inova 500 instruments in CDCl$_3$ solutions if not otherwise specified. Proton chemical shifts are reported in ppm relative to the residual peak of the deuterated solvent or tetramethylsilane: δ (ppm) = 0 for tetramethylsilane, 2.49 for [D$_5$]DMSO, 7.26 for CHCl$_3$. For the characterization of the observed signal multiplicities the following abbreviations were applied: s =

singlet, d = doublet, t = triplet, q = quartet, quin = quintet, m = multiplet, as well as br = broad; J in Hz. ^{13}C chemical shifts are reported relative to the solvent peak or tetramethylsilane: 0 for tetramethylsilane, 39.5 for [D$_5$]DMSO, 77.0 for CDCl$_3$.

IR: Bruker IFS 66 (FT-IR) spectrometer, measured as KBr pellets or oils between KBr plates.

MS: EI-MS: Finnigan MAT 95, 70 eV, DCI-MS: Finnigan MAT 95, 200 eV, reactant gas NH$_3$; ESI-MS: Finnigan LCQ. High resolution mass spectrometry (HRMS): APEX IV 7T FTICR, Bruker Daltonic.

Melting points: Büchi 540 capillary melting point apparatus, uncorrected values.

Elemental analyses: Mikroanalytisches Laboratorium des Instituts für Organische und Biomolekulare Chemie der Universität Göttingen.

Experimental Procedures for the Compounds Described in Chapter 1 "Oligosubstituted Pyrroles Directly from Substituted Methyl Isocyanides and Acetylenes"

Methyl 4-Methoxy-pent-2-ynoate (168b)

To a solution of 3-methoxy butyn-1 (4.45 g, 53 mmol) in anhydrous diethyl ether (200 mL) kept under nitrogen was added dropwise with magnetic stirring at −78 °C 2.5 M solution of n-BuLi (22 mL, 55 mmol). After stirring at −78 °C for 30 min, the mixture was warmed to 0 °C and methyl chloroformate (5.48 g, 58 mmol) was added. The mixture was stirred for 3h at r.t. and quenched with saturated solution of ammonium chloride (80 mL). The organic and water phases were separated and the water phase was extracted with diethyl ether (3 × 50 mL). The combined organic phase was dried over anhydrous Na_2SO_4, filtrated and solvents were removed under reduced pressure. The residue was distilled (10 Torr, b.p 85–90 °C) to give 6.09 g (81%) of product as colorless oil ^{1}H NMR (300 MHz, $CDCl_3$, 25 °C, TMS) δ = 4.18 (q, J = 4.5 Hz, 1 H, CH), 3.79 (s, 3 H, CO_2CH_3), 3.42 (s, 3 H, OCH_3), 1.47 ppm (d, J = 7.8 Hz, 3 H, CH_3); ^{13}C NMR (75.5 MHz, $CDCl_3$, 25 °C): δ = 153.5 (C), 86.8 (C), 76.5 (C), 66.4 (CH), 56.7 (CH_3), 52.6 (CH_3), 20.9 ppm (CH_3); MS (EI) m/z (%): 142.1 (30) [M$^+$], 99.1(28), 59.1(40), 43.1 (100); IR (KBr): 2992, 2940, 2826, 2239, 1727, 1436, 1256, 1115, 1076, 1044, 996, 913, 752, 623 cm^{-1}.

Methyl 3-(4-ethoxyphenyl)propiolate (168d)

Methyl 3-(4-ethoxyphenyl)propiolate (8.98 g, 88%) was prepared analog to methyl 4-methoxy-pent-2-ynoate from 4-ethoxyphenyl acetylene (7.3 g, 50 mmol) and methyl chloroformate (5.2 g, 55 mmol), after recrystallization from hexane/benzene as colorless solid, m.p. 55 °C. ^{1}H NMR (300 MHz, $CDCl_3$, 25 °C, TMS) δ = 7.52 (d, J = 9.0 Hz, 2 H, Ar-H), 6.87 (d, J = 9.0 Hz, 1 H, Ar-H), 4.05 (q, J = 7.3 Hz, 2 H, CH_2), 3.82 (s, 3 H, CO_2CH_3), 1.42 ppm (t, J = 7.3 Hz, 3 H, CH_3); ^{13}C NMR (75.5 MHz, $CDCl_3$, 25 °C): δ = 160.9 (C), 154.7 (C), 134.9 (CH), 114.7 (CH), 111.0 (C), 87.4 (C), 79.7 (C), 63.6 (CH_2), 52.6 (CH_3), 14.6 ppm (CH_3); MS (EI) m/z (%): 204.1 (100) [M$^+$], 173.1(28), 145.1(75), 118.1 (164); IR (KBr): 2361, 2339, 2216, 1700, 1653, 1507, 1288, 1255, 1199, 1164, 1114, 1040, 922, 884, 830, 807, 745 cm^{-1}; elemental analysis calcd (%) for $C_{12}H_{12}O_3$: C 70.67, H 5.92; found: C 71.03, H 6.01.

General Procedure for the Formal Cycloaddition of Substituted Methyl Isocyanides to Acetylenes Mediated by Potassium *tert*-Butoxide (GP1, Method A)

To a solution of the respective acetylene **168** (5.0 mmol) and the respective substituted methyl isocyanide **63** (5.5 mmol) in THF (60 mL) was added dropwise at 20 °C within 1 h a solution of KO*t*Bu (616 mg, 5.5 mmol) in THF (35 mL). The mixture was stirred at 20 °C for 1 h, the reaction then quenched with glacial AcOH (1 mL), and the solution concentrated under reduced pressure. The residue was triturated with CH_2Cl_2 (3 × 30 mL) at 20 °C to extract the crude product, which was purified by column chromatography.

General Procedure for the Copper-Catalyzed Formal Cycloaddition of Substituted Methyl Isocyanides to Acetylenes (GP2, Method B)

The copper catalyst [preferably preactivated nanosize copper powder (3 mg, 0.05 mmol, 5 mol %), or copper thiophenolate (9 mg, 0.05 mmol, 5 mol %)] was added to a solution of the respective substituted methyl isocyanide **63** (1.1 mmol) and the respective acetylene **168** (1.0 mmol) in DMF (2 mL), and the mixture was vigorously stirred at 85 °C (if not otherwise specified) for 16 h. The solvent was removed in vacuo (0.05 mbar), and the residue was purified by column chromatography to give the corresponding pyrrole.

Dimethyl 3-Cyclopropyl-1*H*-pyrrole-2,4-dicarboxylate (173aa)

Following GP2 (Method B), the pyrrole **173aa** (1.03 g, 93%) was obtained from methyl cyclopropylpropiolate (**168a**) (620 mg, 5.0 mmol) and methyl isocyanoacetate (**25-Me**) (545 mg, 5.5 mmol), after column chromatography (cyclohexane/ethyl acetate 4 : 1) as a colorless solid, m. p. 123 °C. ^{1}H NMR (300 MHz, $CDCl_3$, 25 °C, TMS): δ = 9.78 (br s, 1 H, NH), 7.43 (d, *J* = 3.6 Hz, 1 H, NCH), 3.82 (s, 3 H, CH_3), 3.76 (s, 3 H, CH_3), 2.27–2.17 (m, 1 H, cPr-H), 0.96–0.83 ppm (m, 4 H, CH_2); ^{13}C NMR (75.5 MHz, $CDCl_3$, 25 °C): δ = 164.5 (C), 161.4 (C), 135.4 (C), 127.5 (CH), 121.3 (C), 116.9 (C), 51.5 (CH_3), 51.0 (CH_3), 8.2 (CH_2), 7.3 ppm (CH); IR (KBr): 3325 cm^{-1}, 3146, 3010, 2951, 1719, 1696, 1541, 1437, 1276, 1199, 1059, 785; MS (EI): *m/z* (%): 223.1 [M$^+$]; HRMS (ESI): *m/z* calcd for $C_{11}H_{14}NO_4^+$ [M+H$^+$]: 224.0923; found: 224.0917.

Dimethyl 3-(Thiophen-2-yl)-1*H*-pyrrole-2,4-dicarboxylate (173ag)

The pyrrole **173ag** (250 mg, 94%) was obtained from methyl (thiophen-2-yl)propiolate (**168g**) (166 mg, 1.0 mmol) and methyl isocyanoacetate (**25-Me**) (149 mg, 1.5 mmol) following GP2 (Method B) with *NP* Cu^0 (3 mg, 0.05 mmol, 5 mol %) at 120 °C, after column chromatography (hexane/ethyl acetate 2 : 1, R_f = 0.20) as a yellow solid, m.p. 146 °C. ^{1}H NMR (300 MHz, CDCl$_3$, 25 °C, TMS): δ = 9.50 (br s, 1 H, NH), 7.56 (d, *J* = 3.3 Hz, 1 H, NCH), 7.38 (t, *J* = 3.3 Hz, 1 H, thienyl-5H), 7.05 (d, *J* = 3.0 Hz, 2 H, thienyl-3,4H), 3.73 (s, 3 H, CH$_3$), 3.70 ppm (s, 3 H, CH$_3$); ^{13}C NMR (75.5 MHz, CDCl$_3$, 25 °C): δ = 163.7 (C), 160.9 (C), 132.9 (C), 128.5 (CH), 126.9 (CH), 126.2 (CH), 126.0 (CH), 124.1 (C), 121.9 (C), 117.8 (C), 51.8 (CH$_3$), 51.2 ppm (CH$_3$); IR (KBr): 2954, 1731, 1703, 1524, 1439, 1386, 1264, 1197, 1015, 921, 784, 689 cm^{-1}; MS (EI): *m/z* (%): 265.2 (90) [M$^+$], 233.1 (78), 202.1 (62), 43.1 (100); elemental analysis calcd (%) for C$_{12}$H$_{11}$NO$_4$S: C 54.33, H 4.18, N 5.28; found: C 54.05, H 4.10, N 5.38.

Dimethyl 3-(4-Ethoxyphenyl)-1*H*-pyrrole-2,4-dicarboxylate (173ac)

The pyrrole **173ac** (226 mg, 75%) was obtained from methyl (4-ethoxyphenyl)propiolate (**168c**) (204 mg, 1.0 mmol) and methyl isocyanoacetate (**25-Me**) (109 mg, 1.1 mmol) following GP2 (Method B) with *NP* Cu^0 (3 mg, 0.05 mmol, 5 mol %) at 120 °C, after column chromatography (hexane/ethyl acetate 2 : 1, R_f = 0.15) as a colorless solid, m.p. 136 °C. ^{1}H NMR (300 MHz, CDCl$_3$, 25 °C, TMS): δ = 9.73 (br s, 1 H, NH), 7.53 (d, *J* = 3.5 Hz, 1 H, NCH), 7.27 (d, *J* = 8.7 Hz, 2 H, Ar), 6.90 (d, *J* = 8.7 Hz, 2 H, Ar), 4.07 (q, *J* = 7.4 Hz, 2 H, CH$_2$), 3.68 (s, 3 H, CO$_2$CH$_3$), 1.43 ppm (t, *J* = 6.7 Hz, 3 H, CH$_3$); ^{13}C NMR (75.5 MHz, CDCl$_3$, 25 °C): δ = 164.2 (C), 161.3 (C), 158.2 (C), 132.5 (C), 131.3 (2 CH), 127.2 (C), 125.0 (C), 120.4 (CH), 116.4 (C), 113.1 (2 CH), 63.2 (CH$_2$), 51.5 (CH$_3$), 51.0 (CH$_3$), 14.9 ppm (CH$_3$); IR (KBr): 2989, 1732, 1695, 1522, 1436, 1388, 1264, 1006, 922, 829, 785, 522 cm^{-1}; MS (EI): *m/z* (%): 303.2 (100) [M$^+$], 271.2 (44), 243.2 (30), 212.1 (26); elemental analysis calcd (%) for C$_{16}$H$_{17}$NO$_5$: 63.36, H 5.65, N 4.62; found: C 63.18, 5.53, 4.50.

Dimethyl 3-(4-Fluorophenyl)-1*H*-pyrrole-2,4-dicarboxylate (173ad)

The pyrrole **173ad** (215 mg, 78%) was obtained from methyl (4-fluorophenyl)propiolate (**168d**) (178 mg, 1.0 mmol) and methyl isocyanoacetate (**25**-Me) (149 mg, 1.5 mmol) following GP2 (Method B) with *NP* Cu^0 (3 mg, 0.05 mmol, 5 mol %) at 120 °C, after column chromatography (hexane/ethyl acetate 2 : 1, R_f = 0.20) as a colorless solid, m.p. 174 °C. ^{1}H NMR (300 MHz, $CDCl_3$, 25 °C, TMS): δ = 9.56 (br s, 1 H, NH), 7.57 (d, *J* = 3.5 Hz, 1 H, NCH), 7.38–7.26 (m, 2 H, Ar), 7.11–6.98 (m, 2 H, Ar), 3.69 (s, 3 H, CH_3), 3.68 ppm (s, 3 H, CH_3); ^{13}C NMR (75.5 MHz, $CDCl_3$, 25 °C): δ = 164.0 (C), 163.9 (C), 161.1 (C), 131.9 (CH), 131.8 (CH), 131.5 (C), 129.0 (C), 127.1 (CH), 120.7 (C), 116.7 (C), 114.3 (CH), 114.0 (CH), 51.6 (CH_3), 51.1 ppm (CH_3); IR (KBr): 3002, 2955, 1702, 1695, 1598, 1522, 1435, 1386, 1262, 1164, 1098, 1002, 834, 816, 783, 716, 602, 518 cm^{-1}; MS (EI): *m/z* (%): 277.2 (100) [M^+], 245.2 (44), 214.1 (92); elemental analysis calcd (%) for $C_{14}H_{12}FNO_4$: 60.65, H 4.36, N 5.05; found: 60.37, 4.24, 5.06.

Dimethyl 3-(4-(Trifluoromethyl)phenyl)-1*H*-pyrrole-2,4-dicarboxylate (173ae)

The pyrrole **173ae** (341 mg, 70%) was obtained from methyl (4-trifluoromethylphenyl)propiolate (**168e**) (340 mg, 1.5 mmol) and methyl isocyanoacetate (**25**-Me) (254 mg, 2.4 mmol) following GP2 (Method B) with *NP* Cu^0 (5 mg, 0.08 mmol, 5.5 mol %) at 120 °C, after column chromatography (hexane/ethyl acetate 2 : 1 to 1 : 1, R_f = 0.63, hexane/ethyl acetate 1 : 1) as a yellow solid, m.p. 156–157. ^{1}H NMR (300 MHz, DMSO-d_6, 25 °C, TMS): δ = 12.50–12.60 (br s, 1 H, NH), 7.64 (d, *J* = 8.1 Hz, 2 H, Ar-CH), 7.63–7.64 (m, 1 H, NCH), 7.47 (d, *J* = 8.1 Hz, 2 H, Ar-CH), 361 (s, 3 H, CH_3), 3.57 ppm (s, 3 H, CH_3); ^{13}C NMR (75.5 MHz, DMSO-d_6, 25 °C): δ = 163.2 (C), 160.1 (C), 138.3 (C), 131.1 (CH), 130.2 (CH), 127.9 (C), 127.4 (C, q, J_{C-F} = 31.4 Hz), 125.6 (C), 123.5 (C, m, CF_3), 120.4 (C), 115.0 (CH), 51.1 (CH_3), 50.6 ppm (CH_3); IR (KBr): 3300, 1700, 1617 1559, 1522, 1437, 1393, 1322, 1264, 1192, 1172, 1128, 1065, 1019, 847, 786 cm^{-1}; MS (ESI): *m/z* (%): 350 (76) [M + Na^+], 328 (100) [M + H^+]; elemental analysis calcd (%) for $C_{15}H_{12}F_3NO_4$: 55.05, H 3.70, N 4.28; found: C 55.10, H 3.82, N 4.15.

Dimethyl 3-(Pyridin-2-yl)-1H-pyrrole-2,4-dicarboxylate (173af)

The pyrrole **173af** (177 mg, 68%) was obtained from methyl (pyridin-2-yl)propiolate (**168f**) (161 mg, 1.0 mmol) and methyl isocyanoacetate (**25**-Me) (149 mg, 1.5 mmol) following GP2 (Method B) with *NP* Cu^0 (3 mg, 0.05 mmol, 5 mol %) at 120 °C, after column chromatography (ethyl acetate, R_f = 0.30) as a colorless solid. ^{1}H NMR (300 MHz, CDCl$_3$, 25 °C, TMS): δ = 10.87 (br s, 1 H, NH), 7.78 (m, 2 H, Ar-H), 7.44 (m, 3 H, Ar-H), 3.64 (s, 3 H, CH$_3$), 3.59 ppm (s, 3 H, CH$_3$); ^{13}C NMR (75.5 MHz, CDCl$_3$, 25 °C): δ = 163.9 (C), 160.7 (C), 153.4 (C), 147.9 (C), 135.5 (CH), 131.0 (C), 127.0 (2 CH), 122.8 (C), 121.5 (CH), 116.6 (CH), 51.5 (CH$_3$), 51.1 ppm (CH$_3$); IR (KBr): 3446, 1653(br), 902, 726 cm^{-1}; MS (EI): *m/z* (%): 260.2 (52) [M$^+$], 202.2 (100), 197.1 (94), 171.1 (72), 144.2 (76), 44 (84); elemental analysis calcd (%) for C$_{12}$H$_{11}$NO$_4$S: 54.33, H 4.18, N 5.28; found: 54.45, 4.20, 5.21.

Dimethyl 3-(1-Methoxyethyl)-1H-pyrrole-2,4-dicarboxylate (173ab)

The pyrrole **173ab** (130 mg, 54%) was obtained from methyl 4-methoxypent-2-ynoate (**168b**) (340 mg, 1.0 mmol) and methyl isocyanoacetate **25**-Me (109 mg, 1.1 mmol) following GP2 (Method B) with *NP* Cu^0 (3 mg, 0.05 mmol, 5 mol %) after column chromatography (hexane/ethyl acetate 2 : 1 to 1 : 1, R_f = 0.58, hexane/ethyl acetate 1 : 1) as a yellow solid, m.p. 109–110. ^{1}H NMR (300 MHz, CDCl$_3$, 25 °C, TMS): δ = 9.20–9.40 (br s, 1 H, NH), 7.47 (d, *J* = 3.4 Hz, 1 H, NCH), 5.37 (q, *J* = 6.6 Hz, 1 H, C*H*OMe), 3.87 (s, 3 H, CH$_3$), 3.80 (s, 3 H, CH$_3$), 3.20 (s, 3 H, OCH$_3$), 1.61 (d, *J* = 6.6 Hz, 3 H, C*H$_3$*CH) ppm; ^{13}C NMR (75.5 MHz, CDCl$_3$, 25 °C): δ = 164.2 (C), 161.0 (C), 133.8 (C), 127.4 (CH), 121.0 (C), 116.4 (CH), 71.0 (CH), 56.8 (CH$_3$), 51.8 (CH$_3$), 51.3 (CH$_3$), 20.6 (CH$_3$) ppm; IR (KBr): 3383, 1700, 1559, 1506, 1437, 1403, 1340, 1269, 1197, 1080, 1024, 988, 788, 731 cm^{-1}; MS (ESI): *m/z* (%): 505 (55) [2M + Na$^+$], 264 (100) [M + Na$^+$], 242 (12) [M + H$^+$]; HRMS (ESI): calcd for C$_{11}$H$_{15}$NNaO$_5^+$ [M+Na$^+$]: 264.08424; found: 264.08434.

Methyl 2-(Ethoxycarbonyl)-3-cyclopropyl-1*H*-pyrrole-4-carboxylate (173ba)

The pyrrole **173ba** (422 mg, 89%) was obtained from methyl cyclopropylpropiolate (**168a**) (248 mg, 2.0 mmol) and ethyl isocyanoacetate (**25**-Et) (249 mg, 2.2 mmol) following GP1 (Method A), after column chromatography (hexane/ethyl acetate 5 : 1) as a yellow oil, which crystallized to form a yellow solid, m.p. 45–46 °C. Alternatively, this compound (413 mg, 89%) can be synthesized following GP2 (Method B). ^{1}H NMR (300 MHz, CDCl$_3$, 25 °C, TMS): δ = 9.56 (br s, 1 H, NH), 7.45 (d, *J* = 3.8 Hz, 1 H, NCH), 4.35 (q, *J* = 7.2 Hz, 2 H, Et-CH$_2$), 3.81 (s, 3 H, CO$_2$CH$_3$), 2.29–2.20 (m, 1 H, CH), 1.38 ppm (t, *J* = 7.2 Hz, 3 H, Et-CH$_3$); ^{13}C NMR (75.5 MHz, CDCl$_3$, 25 °C): δ = 164.4 (C), 161.0 (C), 135.1 (C), 127.1 (CH), 121.8 (C), 117.1 (C), 60.6 (CH$_3$), 51.0 (CH$_2$), 14.3 (CH), 8.4 (CH$_3$), 7.3 ppm (CH$_2$) ; IR (KBr): 3301, 2985, 1693, 1551, 1413, 1270, 1193, 1103, 1025, 931, 782 cm^{-1}; MS (EI): *m/z* (%): 237.2 [M$^+$] (56), 205.1 (52), 176.1 (53), 132.1 (100); elemental analysis calcd (%) for C$_{12}$H$_{15}$NO$_4$: C 60.75, H 6.37, N 5.90; found: C 60.81, H 6.28, N 6.09.

Methyl 2-(Ethoxycarbonyl)-1*H*-pyrrole-4-carboxylate (173bh)[181]

The pyrrole **173bh** (73 mg, 37%) was obtained from 84 mg (1.0 mmol) of methyl propiolate (**168h**) and 113 mg (1.0 mmol) of ethyl isocyanoacetate (**25**-Et) following GP2 (Method B) with *NP* Cu^0 (3 mg, 0.05 mmol, 5 mol %) at 60 °C, after column chromatography (hexane/ethyl acetate 4 : 1, *R*$_f$ = 0.17) as a colorless solid, m.p. 98–99 °C. ^{1}H NMR (300 MHz, CDCl$_3$, 25 °C, TMS) δ = 9.98–9.76 (br s, 1 H, NH), 7.55 (dd, *J* = 3.2, 1.5 Hz, 1 H, CH) 7.31 (dd, *J* = 2.4, 1.6 Hz, 1 H, NCH), 4.35 (q, *J* = 7.0 Hz, 2 H, CH$_2$), 3.84 (s, 3 H, CO$_2$CH$_3$), 1.37 ppm (t, *J* = 7.2 Hz, 3 H, Et-CH$_3$); ^{13}C NMR (75.5 MHz, CDCl$_3$, 25 °C): δ = 164.4 (C), 161.0 (C), 127.0 (C), 123.8 (CH), 117.8 (C), 115.8 (CH), 60.9 (CH$_2$), 51.3 (CH$_3$), 14.3 ppm (CH$_3$); MS (EI) *m/z* (%): 197.0 (76) [M$^+$], 166.1(53), 152.1(44), 120.1(100); IR (KBr): 3293, 2981, 1690, 1562, 1499, 1441, 1403, 1280, 1216, 1122, 1085, 1022, 989, 964, 927, 853, 762, 604, 504 cm^{-1}; elemental analysis calcd (%) for C$_9$H$_{11}$NO$_4$: C 54.82, H 5.62, N 7.10; found: C 54.92, H 5.82, N 6.98.

Methyl 2-(4-Toluenesulfonyl)-1H-pyrrole-4-carboxylate (173ch)[182]

The pyrrole **173ch** (83 mg, 30%) was obtained from 84 mg (1.0 mmol) of methyl propiolate (**168h**) and 195 mg (1.0 mmol) of tosylmethyl isocyanide (**41**-H) following GP2 (Method B) with NP Cu^0 (3 mg, 0.05 mmol, 5 mol %) at 60 °C, after column chromatography (hexane/ethyl acetate 2 : 1, R_f = 0.22) as a colorless solid, m.p. 157–158 °C. Alternatively, it was prepared with KOtBu as a mediator (105 mg, 38%). ^{1}H NMR (300 MHz, CDCl$_3$, 25 °C, TMS): δ = 9.95 (br s, 1 H, NH), 7.81 (d, J = 8.1 Hz, 2 H, Ts-CH), 7.53 (dd, J = 3.1, 1.6 Hz, 1 H, CH), 7.30 (d, J = 8.1 Hz, 2 H, Ts-CH), 7.21 (dd, J = 3.1, 1.6 Hz, 1 H, NCH), 3.80 (s, 3 H, CO$_2$CH$_3$), 2.41 ppm (s, 3 H, CH$_3$); ^{13}C NMR (75.5 MHz, CDCl$_3$, 25 °C): δ = 163.8 (C), 144.6 (CH), 138.4 (C), 130.2 (C), 130.0 (2 CH), 127.3 (C), 127.1 (2 CH), 118.5 (C), 115.7 (CH), 51.6 (CH$_3$), 21.6 ppm (CH$_3$); MS (ESI): m/z (%): 302.0 [M+Na$^+$], 278.3 [M-H$^-$]; IR (KBr): 3250, 2950, 1691, 1595, 1546, 1473, 1433, 1392, 1319, 1228, 1183, 1145, 1116, 1076, 1017, 988, 930, 857, 813, 766, 744, 706, 676, 623, 604, 535, 492 cm^{-1}; HRMS (ESI): calcd for C$_{13}$H$_{14}$NO$_4$S$^+$ [M+H]$^+$: 280.06381; found: 280.06403.

Methyl 2-Phenyl-1H-pyrrole-4-carboxylate (173eh)[183]

The pyrrole **173eh** (47 mg, 25%) was obtained from 84 mg (1.0 mmol) of methyl propiolate (**168h**) and 102 mg (1.0 mmol) of phenylmethyl isocyanide (**63e**) following GP2 (Method B) with NP Cu^0 (3 mg, 0.05 mmol, 5 mol %) at 60 °C, after column chromatography (hexane/ethyl acetate 4 : 1, R_f = 0.19) as a colorless solid, m.p. 163–164 °C. Alternatively, it was prepared with KOtBu as a mediator (13 mg, 7%). ^{1}H NMR (300 MHz, CDCl$_3$, 25 °C, TMS): δ = 8.90 (br s, 1 H, NH), 7.51 – 7.47 (m, 3 H, Ph), 3.39 (t, J = 2.8 Hz, 2 H, Ph), 7.26 (t, J = 2.8 Hz, 1 H, CH), 6.92 (m, 1 H, NCH), 3.84 ppm (s, 3 H, CH$_3$); ^{13}C NMR (75.5 MHz, CDCl$_3$, 25 °C): δ = 165.4 (C), 133.1 (C), 131.7 (C), 129.0 (CH), 127.1 (CH), 124.2 (CH), 124.1 (CH), 117.7 (C), 106.6 (C), 51.2 ppm (CH$_3$); MS (EI): m/z (%): 201.1(86) [M$^+$], 170.1(100); IR (KBr): 3289, 3019, 1679, 1604, 1572, 1515, 1491, 1441, 1401, 1352, 1283, 1222, 1145, 1132, 996, 928, 833, 808, 767, 725, 693, 659, 605, 524, 505 cm^{-1}; HRMS (ESI): calcd for C$_{12}$H$_{12}$NO$_2$$^+$ [M+H]$^+$: 202.08626; found: 202.08629.

Methyl (4-Nitrophenyl)-1*H*-pyrrole-4-carboxylate (173fh)

The pyrrole **173fh** (109 mg, 44%) was obtained from 84 mg (1.0 mmol) of methyl propiolate (**168h**) and 162 mg (1.0 mmol) of 4-nitrophenylmethyl isocyanide (**63f**) following GP2 (Method B) with NP Cu^0 (3 mg, 0.05 mmol, 5 mol %) at 60 °C, after column chromatography (hexane/ethyl acetate 2 : 1, R_f = 0.30) as a yellow solid, m.p. 225 °C. ^{1}H NMR (300 MHz, [d6]DMSO, 25 °C): δ = 12.29 (br s, 1 H, NH), 8.19 (d, J = 9.0 Hz, 2 H, Ar-CH), 7.93 (d, J = 8.7 Hz, 2 H, Ar-CH), 7.65 (dd, J = 3.1, 1.6 Hz, 1 H, CH), 7.18 (dd, J = 3.1, 1.6 Hz, 1 H, NCH), 3.74 ppm (s, 3 H, CH$_3$); ^{13}C NMR (75.5 MHz, [d6]DMSO, 25 °C): δ = 164.0 (C), 145.1 (C), 137.9 (C), 130.5 (C), 127.0 (CH), 124.2 (CH), 124.1 (CH), 116.9 (C), 109.7 (CH), 50.7 ppm (CH$_3$); MS (EI): m/z (%): 246.1 (100) [M$^+$], 215.1 (80); IR (KBr): 3268, 3121, 1674, 1599, 1509, 1486, 1449, 1433, 1391, 1335, 1288, 1228, 1140, 1108, 991, 928, 851, 818, 769, 752, 691, 602, 521 cm^{-1}; elemental analysis calcd (%) for C$_{12}$H$_{10}$N$_2$O$_4$: C 58.54, H 4.09, N 11.38; found: C 58.26, H 3.92, N 11.00.

General Procedure A for the Synthesis of 2,3-Disubstituted Pyrroles 178 (GP3)

An oven-dried Schlenk flask equipped with magnetic stirrer and rubber septum, was charged with CuBr (143.5 mg, 1.0 mmol), Cs$_2$CO$_3$ (326 mg, 1 mmol) and DMF (5 mL), evacuated and refilled with nitrogen. The respective acetylene **167** (1.0 mmol) was added from a syringe with stirring, and the mixture was heated at 120 °C for 10 min, then a solution of the respective isocyanide **63** (2.0 mmol) in DMF (5 mL) was injected over a period of 2 h, after that the reaction mixture was stirred at 120 °C for another 1 h. After cooling and evaporation of the solvent in vacuo, the residue was purified by column chromatography on silica gel (eluting with 5 : 1 to 1 : 1 hexane/ethyl acetate) to provide the desired product.

General Procedure B for the Synthesis of 2,3-Disubstituted Pyrroles 178 (GP4)

An oven-dried Schlenk flask equipped with magnetic stirrer and rubber septum was charged with CuBr (143.5 mg, 1.0 mmol), Cs$_2$CO$_3$ (326 mg, 1 mmol) and DMF (5 mL), evacuated and refilled with nitrogen. The respective acetylene **167** (1.0 mmol) was added from a syringe with stirring, and the mixture was heated at 120 °C for 10 min, then solutions of the respective isocyanide **63** (1.0 mmol) and the respective acetylene **167** (1.0 mmol) in DMF (5 mL) were injected over a period of 2 h, after that the reaction mixture was stirred at 120 °C for another 1 h. After cooling and evaporation of the solvent in vacuo, the residue was purified by column chromatography on silica gel (eluting with 5 : 1 to 1 : 1 hexane/ethyl acetate) to provide the desired product.

Ethyl 3-Butyl-1-*H*-pyrrol-2-carboxylate (178ba)

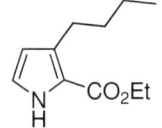

The pyrrole **178ba** (250 mg, 64%) was obtained from 1-hexyne (**167a**) (328 mg, 4 mmol), ethyl isocyanoacetate (**25**-Et) (226 mg, 2 mmol) following GP4, after column chromatography (hexane/ethyl acetate 4 : 1, R_f = 0.45) as a colorless oil. Alternatively it was obtained following GP3 (273 mg, 70%). ^{1}H NMR (300 MHz, CDCl$_3$, 25 °C, TMS): δ = 9.10–8.89 (br m, 1 H, NH), 6.81 (t, *J* = 3 Hz, 1 H, CH), 6.10 (t, *J* = 3 Hz , 1 H, CH), 4.29 (q, *J* = 7 Hz , 2 H, Et-CH$_2$), 2.77 (t, *J* = 8 Hz , 2 H, CH$_2$), 1.60–1.20 (m, 7 H), 0.91 ppm (t, *J* = 8 Hz, 3 H, Et-CH$_3$); ^{13}C NMR (75.5 MHz, CDCl$_3$, 25 °C): δ = 161.7 (C), 133.3 (C), 121.5 (CH), 118.8 (C), 111.4 (CH), 59.9 (CH$_2$), 33.0 (CH$_2$), 26.6 (CH$_2$), 22.6 (CH$_2$), 14.4 (CH$_3$), 13.9 ppm (CH$_3$); IR (film): 3322, 2957, 2860, 1672, 1561, 1420, 1318, 1262, 1188, 1133, 1044, 783 cm^{-1}; MS (EI): m/z (%): 195 (72) [M$^+$], 153 (40), 124 (100), 106 (56), 80 (40); elemental analysis calcd (%) for C$_{11}$H$_{17}$NO$_2$: C 67.66, H 8.78, N 7.17; found: C 67.71, H 8.51, N 7.02.

Ethyl 3-(Methoxymethyl)-1*H*-pyrrole-2-carboxylate (178bb)

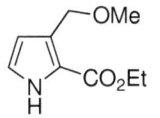

The pyrrole **178bb** (88 mg, 48%) was obtained from 226 mg (2.0 mmol) of ethyl isocyanoacetate (**25**-Et) and 70 mg (1.0 mmol) of 3-methoxypropyne (**167b**) following GP3 as a colorless solid, m.p. 74 °C. R_f = 0.27 (hexane/ethylacetate 4 : 1). Alternatively, it was prepared following GP4 (83 mg, 45%). ^{1}H NMR (300 MHz, CDCl$_3$, 25 °C, TMS): δ = 9.39 (br s, 1 H, NH), 6.88 (t, *J* = 2.6 Hz, 1 H, CH), 6.34 (t, *J* = 2.6 Hz, 1 H, CH), 4.69 (s, 2H, CH$_2$), 4.34 (q, *J* = 7.2 Hz, 2 H, Et-CH$_2$), 3.43 (s, 3 H, OCH$_3$), 1.37 ppm (t, *J* = 7.2 Hz, 3H, CH$_3$); ^{13}C NMR (75 MHz, CDCl$_3$, 25 °C, TMS): δ = 161.2 (C), 128.4 (C), 121.9 (CH), 119.1 (C), 111.1 (CH), 67.2 (CH$_2$), 60.3 (CH$_3$), 58.1 (CH$_2$), 14.4 ppm (CH$_3$); MS (EI): *m/z* (%): 183.2 (40) [M$^+$], 168.1 (52), 154.2 (45), 122.1 (100); IR (KBr): 3288, 1671, 1490, 1426, 1373, 1326, 1271, 1222, 1193, 1139, 1111, 963, 782, 751, 601 cm^{-1}; elemental analysis calcd (%) for C$_9$H$_{13}$NO$_3$: C 59.00, H 7.15, N 7.65; found: C 59.06, H 6.80, N 7.35.

Ethyl 3-Cyclopropyl-1H-pyrrole-2-carboxylate (178be)

The pyrrole **178be** (157 mg, 88%) was obtained following GP3 from 226 mg (2.0 mmol) of ethyl isocyanoacetate (**25**-Et) and 66 mg (1.0 mmol) of cyclopropylacetylene (**167e**) as a colorless solid, m.p. 51–52 °C, R_f = 0.37 (hexane/ethyl acetate 5 : 1). ^{1}H (300 MHz, CDCl$_3$, 25 °C, TMS): δ = 8.92 (br s, 1 H, NH), 6.79 (t, J = 2.9 Hz, 1 H, CH), 5.78 (t, J = 2.9 Hz, 1 H, CH), 4.35 (q, J = 7.2 Hz, 2 H, CH$_2$), 2.57–2.48 (m, 1 H, cPr-CH), 1.37 (t, J = 7.2 Hz, 3 H, CH$_3$), 0.99–0.93 (m, 2 H, cPr-CH$_2$), 0.64–0.59 ppm (m, 2 H, cPr-CH$_2$); ^{13}C (75.5 MHz, CDCl$_3$, 25 °C): δ = 161.7 (C), 135.5 (C), 121.9 (CH), 119.7 (C), 106.2 (CH), 60.0 (CH$_2$), 14.5 (CH), 9.3 (CH$_2$), 7.9 ppm (CH$_2$); MS (EI) m/z (%): 179.2 (100) [M$^+$], 150.2 (45), 133.2 (39), 106.2 (62); IR (KBr): 3299, 1673, 1422, 1391, 1322, 1279, 1218, 1185, 1141, 1036, 907, 781, 745, 602 cm^{-1}; elemental analysis calcd (%) for C$_{10}$H$_{13}$NO$_2$: C 67.02, H 7.31, N 7.82; found: C 67.66, H 6.80, N 7.36.

Ethyl 3-tert-Butyl-1H-pyrrole-2-carboxylate (178bf) and Ethyl 4-tert-butyl-1H-pyrrole-2-carboxylate (iso-178bf)[184]

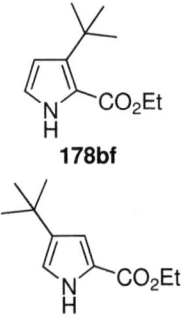

178bf

iso-178bf

A 5 : 1 mixture of the regioisomeric pyrroles **178bf** and *iso*-**178bf** (10 mg, 5%) was obtained following GP4 from 113 mg (1.0 mmol) of ethyl isocyanoacetate (**25**-Et) and 164 mg (2.0 mmol) of *tert*-butylacetylene (**167f**), as a colorless oil, R_f = 0.43 (hexane/ethyl acetate 4 : 1). **178bf**: ^{1}H (300 MHz, CDCl$_3$, 25 °C, TMS): δ = 9.16 (br s, 1 H, NH), 6.78 (t, J = 2.6 Hz, 1 H, CH), 6.21 (t, J = 2.6 Hz, 1 H, CH), 4.32 (q, J = 7.2 Hz, 2 H, CH$_2$), 1.40 (s, 9 H, *t*Bu), 1.25 ppm (s, 3 H, CH$_3$); ^{13}C (75.5 MHz, CDCl$_3$, 25 °C): δ = 160.4 (C), 142.6 (C), 120.0 (CH), 109.8 (CH), 109.2 (C), 60.0 (CH$_2$), 31.6 (CH$_3$), 30.2 (CH$_3$), 22.6 ppm (C); *iso*-**178bf**: ^{1}H (300 MHz, CDCl$_3$, 25 °C, TMS): δ = 8.95 (br s, 1 H, NH), 6.83 (t, J = 2.6 Hz, 1 H, CH), 6.12 (t, J = 2.6 Hz, 1 H, CH), 4.31 (q, J = 7.2 Hz, 2 H, CH$_2$), 1.37 (t, J = 7.2 Hz, 9 H, *t*Bu), 0.93 ppm (t, J = 7.2 Hz, 3 H, CH$_3$); ^{13}C (75.5 MHz, CDCl$_3$, 25 °C): δ = 160.4 (C), 142.6 (C), 121.4 (C), 117.9 (CH), 111.4 (CH), 59.9 (CH$_2$), 33.0 (CH$_3$), 29.7 (CH$_3$), 26.6 ppm (C); MS (EI) m/z (%): 195.2 (26) [M$^+$], 180.2 (28), 134.2 (100).

4,5-Dihydro-1-*H*-pyrano[3,4-*b*]pyrrol-7-one (179)

The δ-lactone-annelated pyrrole **179** (51 mg, 37%) was obtained from 113 mg (1.0 mmol) of ethyl isocyanoacetate (**25**-Et) and 140 mg (2.0 mmol) of but-3-yn-1-ol (**167i**) following GP4, as a colorless solid, m.p. 123–124 °C. Alternatively, **179** was prepared in 44% yield following GP3, R_f = 0.45 (hexane/ethyl acetate 1 : 1). ^{1}H NMR (300 MHz, CDCl$_3$, 25 °C, TMS): δ = 10.68 (br s, 1 H, NH), 7.08 (t, *J* = 2.8 Hz, 1 H), 6.13 (t, *J* = 2.8 Hz, 1 H, CH), 4.56 (t, *J* = 6.2 Hz, 2 H, CH$_2$), 2.93 ppm (t, *J* = 6.2 Hz, 2 H, CH$_2$); ^{13}C NMR (75.5 MHz, CDCl$_3$, 25 °C): δ = 161.2 (C), 130.8 (C), 126.4 (CH), 117.9 (C), 107.2 (CH), 69.5 (CH$_2$), 23.0 ppm (CH$_2$); MS (EI) *m/z* (%):137.1 (100) [M$^+$], 107.1 (42), 79.1 (78); IR (KBr): 3274, 1686, 1400, 1308, 1274, 1209, 1185, 1123, 1078, 1049, 1013, 773, 739, 599, 496, 460 cm^{-1}; elemental analysis calcd (%) for C$_7$H$_7$NO$_2$: C 61.31, H 5.14, N 10.21; found: C 61.51, H 4.98, N 10.18.

Ethyl 3-(1-Methoxy-ethyl)-1*H*-pyrrole-2-carboxylate (178bc)

The pyrrole **178bc** (145 mg, 74%) was obtained from 226 mg (2.0 mmol) of ethyl isocyanoacetate (**25**-Et) and 84 mg (1.0 mmol) of 3-methoxy but-1-yne (**167c**) following GP3, as a colorless solid, m.p. 53 °C, R_f = 0.31 (hexane/ethyl acetate 4 : 1), m.p. 53 °C. ^{1}H (300 MHz, CDCl$_3$, 25 °C, TMS): δ = 9.32 (br s, 1 H, NH), 6.90 (t, *J* = 2.6 Hz, 1 H, CH), 6.35 (t, *J* = 2.6 Hz, 1 H, CH), 5.04 (q, *J* = 6.2 Hz, 1 H, OCH), 4.35 (q, *J* = 7.0 Hz, 2 H, CH$_2$), 3.29 (s, 3 H, OCH$_3$), 1.44 (d, *J* = 6.5 Hz, 3 H, CH$_3$), 1.37 ppm (t, *J* = 7.2 Hz, 3 H, Et-CH$_3$); ^{13}C (75.5 MHz, CDCl$_3$, 25 °C): δ = 161.2 (C), 134.4 (C), 122.1 (CH), 118.8 (C), 108.4 (CH), 72.1 (CH), 60.2 (CH$_3$), 56.3 (CH$_2$), 22.9 (CH$_3$), 14.4 ppm (CH$_3$); MS (EI): *m/z* (%): 197.3 (18) [M$^+$], 182.2 (100), 136.2 (50), 122.1 (62); IR (KBr): 3220 cm^{-1}, 2982, 2928, 2822, 1702, 1566, 1481, 1419, 1367, 1337, 1306, 1261, 1208, 1190, 1146, 1072, 1037, 902, 846, 785, 739, 610,572; elemental analysis calcd (%) for C$_{10}$H$_{15}$NO$_3$: C 60.90, H 7.67, N 7.10; found: C 61.02, H 7.45, N 6.95.

Ethyl 3-Phenyl-1*H*-pyrrole-2-carboxylate (178bd)[185]

The pyrrole **178bd** (85 mg, 40%), was obtained following GP3 from 226 mg (2.0 mmol) of ethyl isocyanoacetate (**25**-Et) and 102 mg (1.0 mmol) of phenylacetylene (**167d**), as a colorless solid, m.p. 54–55 °C, R_f = 0.40 (hexane/ethyl acetate 2 : 1). ^{1}H (300 MHz, CDCl$_3$, 25 °C): δ = 9.22 (br s, 1 H, NH), 7.58–7.54 (m, 2 H, Ph), 7.40–7.27 (m, 3 H, Ph), 6.95 (t, *J* = 2.6 Hz, 1 H, CH), 6.36 (t, *J* = 2.6 Hz, 1 H, CH), 4.26 (q, *J* = 7.2 Hz, 2 H, CH$_2$), 1.25 ppm (t, *J* = 6.9 Hz, 3 H, CH$_3$); ^{13}C (75.5 MHz, CDCl$_3$, 25°C): δ = 161.1

(C), 135.1 (C), 132.0 (C), 129.5 (CH), 127.6 (CH), 126.9 (CH), 121.7 (CH), 118.2 (C), 112.5 (CH), 60.3 (CH_2), 14.2 ppm (CH_3); MS (EI): *m/z* (%): 215.2 (90) [M$^+$], 169.2(100); IR (KBr): 3295, 2980, 1668, 1504, 1419, 1358, 1321, 1297, 1213, 1153, 1020, 896, 868, 791, 748, 700, 611 cm^{-1}; elemental analysis calcd (%) for $C_{13}H_{13}NO_2$: C 72.54, H 6.09, N 6.51; found: C 72.32, H 6.20, N 6.33.

Ethyl 3-(Pyridin-2-yl-1)-1*H*-pyrrole-2-carboxylate (178bg)

The pyrrole **178bg** (34 mg, 16 %) was obtained following GP3 from 226 mg (2.0 mmol) of ethyl isocyanoacetate (**25-Et**) and 103 mg (1.0 mmol) of 2-ethynylpyridine (**167g**) as a yellow oil, R_f = 0.33 (hexane/ethyl acetate 2 : 1). ^{1}H NMR (300 MHz, $CDCl_3$, 25 °C, TMS): δ = 9.57 (br s, 1 H, NH), 8.66 (d, *J* = 4.9 Hz, 1 H, CH), 7.87 (d, *J* = 7.9 Hz, 1 H, CH), 7.69 (dt, *J* = 7.5, 1.9 Hz, 1 H, CH), 7.19 (ddd, *J* = 7.2, 4.9, 1.1 Hz, 1 H, CH), 6.95 (t, *J* = 2.6 Hz, 1 H, CH), 6.69 (t, *J* = 2.6 Hz, 1 H, CH), 4.28 (q, *J* = 7.2 Hz, 2 H, CH_2), 1.26 ppm (t, *J* = 7.2 Hz, 3 H, CH_3); ^{13}C NMR (75.5 MHz, $CDCl_3$, 25 °C): δ = 160.7 (C), 153.7 (C), 148.9 (CH), 135.4 (CH), 131.3 (C), 124.9 (CH), 121.8 (CH), 121.6 (CH), 118.9 (C), 112.8 (CH), 60.4 (CH_2), 14.2 ppm (CH_3); MS (EI) *m/z* (%): 216.2 (30) [M$^+$], 170.1 (28), 144.2 (100); IR (KBr): 3121, 2981, 1695, 1593, 1564, 1492, 1409, 1294, 1147, 1073, 1024, 896, 776 cm^{-1}; elemental analysis calcd (%) for $C_{12}H_{12}N_2O_2$: C 66.65, H 5.59, N 12.96; found: C 66.85, H 5.42, N 12.79.

Ethyl 3-(*sec*-Butyl)-1*H*-pyrrole-2-carboxylate (178bh)

The pyrrole **178bh** (114 mg, 58%) was obtained following GP3 from 226 mg (2.0 mmol) of ethyl isocyanoacetate (**25-Et**) and 82 mg (1.0 mmol) of 3-methyl pent-1-yne (**167h**) as a yellow oil, R_f = 0.41 (hexane/ethyl acetate 5 : 1). ^{1}H NMR (300 MHz, $CDCl_3$, 25 °C, TMS): δ = 9.03 (br s, 1 H, NH), 6.85 (t, *J* = 2.6 Hz, 1 H, CH), 6.15 (t, *J* = 2.6 Hz, 1 H, CH), 4.32 (q, *J* = 7.2 Hz, 1 H, CH_2), 4.31 (q, *J* = 7.2 Hz, 1 H, CH_2), 3.42–3.30 (m, 1 H, CH), 1.67–1.46 (m, 2 H, *sec*-Bu-CH_2), 1.36 (t, *J* = 7.2 Hz, 3 H, *sec*-Bu-CH_2CH_3), 1.20 (d, *J* = 6.8 Hz, 3 H, (CH)CH_3), 0.86 ppm (t, *J* = 7.2 Hz, 3 H, Et-CH_3); ^{13}C NMR (75.5 MHz, $CDCl_3$, 25 °C): δ = 161.6 (C), 138.8 (C), 121.6 (CH), 118.4 (C), 108.5 (CH), 59.8 (CH_2), 32.2 (CH), 31.0 (CH_2), 21.3 (CH_3), 14.4 (CH_3), 12.1 ppm (CH_3); MS (EI) *m/z* (%): 195.3 (56) [M$^+$], 166.2 (94), 120.2 (100); IR (film): 3326, 2964, 2932, 2873, 1672, 1557, 1455, 1421, 1371, 1318, 1266, 1193, 1132, 1088, 1045, 901, 784 cm^{-1}; elemental analysis calcd (%) for $C_{11}H_{17}NO_2$: C 67.66, H 8.78, N 7.17; found: C 67.80, H 7.50, N 7.01.

tert-Butyl 3-(*n*-Butyl)-1*H*-pyrrole-2-carboxylate (178ca)

The pyrrole **178ca** (127 mg, 47%) was obtained from 170 mg (1.2 mmol) of *tert*-butyl isocyanoacetate (**25**-*t*Bu) and 196 mg (2.4 mmol) of 1-hexyne (**167a**) following GP4 as a light-brown solid, m.p. 42 °C, R_f = 0.48 (hexane/ethyl acetate 5 : 1). ^{1}H (300 MHz, CDCl$_3$, 25 °C, TMS): δ = 8.95 (br s, 1 H, NH), 6.80 (t, *J* = 2.6 Hz, 1 H, CH), 6.09 (t, *J* = 2.6 Hz, 1 H, CH), 2.75 (dd, *J* = 7.9 Hz, *J* = 7.5 Hz, 2 H, CH$_2$), 1.64–1.52 (m, 2 H, CH$_2$), 1.57 (s, 9 H, *t*Bu), 1.38 (m, 2 H, CH$_2$), 0.93 ppm (t, *J* = 7.2 Hz, 3 H, CH$_3$); ^{13}C (75.5 MHz, CDCl$_3$, 25 °C): δ = 161.3 (C), 132.3 (C), 120.8 (CH), 120.0 (C), 111.3 (CH), 80.51 (C), 33.1 (CH$_2$), 28.5 (CH$_3$), 26.8 (CH$_2$), 22.7 (CH$_2$), 14.1 ppm (CH$_3$); MS (EI) *m/z* (%): 223.3 (22) [M$^+$], 167.2 (38), 125.1 (100), 106.1 (26), 80.1 (22); IR (KBr): 3321, 2958, 1672, 1553, 1415, 1367, 1330, 1266, 1173, 1130, 902, 845, 783, 750, 604 cm^{-1}; elemental analysis calcd (%) for C$_{13}$H$_{21}$NO$_2$: C 69.92, H 9.48, N 6.27; found: C 70.17, H 9.19, N 6.09.

3-Butyl-2-(4-nitrophenyl)-1*H*-pyrrole (178fa)

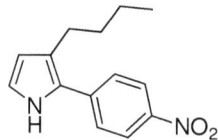

The pyrrole **178fa** (45 mg, 18%) was obtained following GP4 from 162 mg (1.0 mmol) of (4-nitrophenyl)methyl isocyanide (**63f**) and 164 mg (2.0 mmol) of 1-hexyne (**167a**) as a red solid, m.p. 95 °C, R_f = 0.35 (hexane/ethyl acetate 4 : 1). ^{1}H (300 MHz, CDCl$_3$, 25 °C, TMS): δ = 8.27 (br s, 1 H, NH), 8.25 (d, *J* = 9.0 Hz, 2 H, Ar-CH), 7.51 (d, *J* = 9.0 Hz, 2 H, Ar-CH), 6.90 (t, *J* = 2.6 Hz, 1 H, CH), 6.25 (t, *J* = 2.6 Hz, 1 H, CH), 2.68 (dd, *J* = 7.7, 7.5 Hz, 2 H, CH$_2$), 1.68–1.58 (m, 2 H, CH$_2$), 1.47–1.34 (m, 2 H, CH$_2$), 0.93 ppm (t, *J* = 7.2 Hz, 3 H, CH$_3$); ^{13}C (75.5 MHz, CDCl$_3$, 25 °C): δ = 145.2 (C), 140.0 (C), 125.9 (CH), 125.8 (CH), 125.4 (C), 124.3 (CH), 120.0 (C), 111.8 (CH), 33.0 (CH$_2$), 26.6 (CH$_2$), 22.7 (CH$_2$), 14.0 ppm (CH$_3$); MS (EI) *m/z* (%): 244.3 (62) [M$^+$], 201.2 (100), 155.2 (66); IR (KBr): 3374, 2923, 1593, 1496, 1418, 1316, 1172, 1109, 891, 849, 756, 697, 588 cm^{-1}; UV (MeCN): λ$_{max}$ (ε) = 393 (13940), 382 (9428), 197 nm (24045 mol^{-1}dm^{3}cm^{-1}); elemental analysis calcd (%) for C$_{14}$H$_{16}$N$_2$O$_2$: C 68.83, H 6.60, N 11.47; found: C 69.02, H 6.55, N 11.20.

Partly Deuterated Methyl 3-Butyl-1-*H*-pyrrol-2-carboxylate (178aa-D)

(43)57% (D)H [structure: pyrrole with (D)H at position 5 (43)57%, (D)H at position 4 (43)57%, butyl at position 3, CO$_2$Me at position 2, NH]

A mixture of pyrroles **178aa**-D and **178aa**-H (108 mg, 60%) was obtained from 1-deutero hexyne-1 (**167a**-D) (166 mg, 2 mmol), methyl isocyanoacetate (**25**-Me) (99 mg, 1 mmol) following GP4, after column chromatography (hexane/ethyl acetate 4 : 1, R_f = 0.43) as a colorless oil.

^{1}H NMR (300 MHz, CDCl$_3$, 25 °C, TMS): δ = 8.95 (br m, 1 H, NH), 6.83 (t, J = 3 Hz, 0.57 H, CH), 6.13 (t, J = 3 Hz, 0.57 H, CH), 3.84 (s, 3 H, CH$_3$), 2.79 (t, J = 8 Hz, 2 H, CH$_2$), 1.61–1.52 (m, 2 H, CH$_2$), 1.45–1.30 (m, 2 H, CH$_2$), 0.93 ppm (t, J = 8 Hz, 3 H, Et-CH$_3$); MS (EI): m/z (%): 183.2 (18) [M$^+$+2], 182.2 (40) [M$^+$+1], 181.2 (34) [M$^+$].

Experimental Procedures for the Compounds Described in Chapter 2
"ortho-Lithiophenyl Isocyanide: A Versatile Precursor for 3H-Quinazolin-4-ones and 3H-Quinazolin-4-thiones"

N-Formyl-2-iodoaniline

A solution of o-iodoaniline (8.0 g, 36.5 mmol) and ethyl formate (15 mL) in anhydrous THF (250 mL) was added dropwise to a suspension of NaH (60% in mineral oil, 1.82 g, 45.6 mmol) in anhydrous THF (270 mL). The resulting mixture was stirred at r.t. for 24 h, and then the reaction was quenched with cold water (10 mL). The solvents were removed under reduced pressure, and the residue was dissolved in ethyl acetate/water (400/100 mL). The aqueous phase was extracted with ethyl acetate (2 × 100 mL), the combined organic extracts were dried over anhydrous Na_2SO_4 and evaporated. The residue was washed thoroughly with hexane (3 × 100 mL) and dried in vacuo to give 8.84 g (98%) of the title compound as a colorless solid, m. p. 118 °C. R_f = 0.4 (hexane/EtOAc 2 : 1). ^{1}H NMR (300 MHz, DMSO-d6): δ 9.50 (br s, 1 H, NH), 8.34 (s, 1 H, CHO), 7.87 (d, J = 7.7 Hz, 1 H), 7.78 (d, J = 8.1 Hz, 1 H), 7.37 (t, J = 7.3 Hz, 1 H), 6.94 (t, J = 7.3 Hz, 1 H); ^{13}C NMR (75.5 MHz, DMSO-d6): δ 160.1 (CH), 139.0 (CH), 138.4 (C), 128.5 (CH), 126.8 (CH), 126.7 (C), 124.5 (CH); MS (70 eV, EI) m/z (%): 247.1 (48) [M$^+$], 120.1 (100), 92.1 (50), 65.1 (60); IR (KBr): 3223, 2899, 1658 (C=O), 1583, 1572, 1524, 1463, 1435, 1394, 1281, 1240, 1163, 1151, 1017, 885, 746, 693, 644, 520, 461, 429 cm^{-1}; Anal. Calcd for C_7H_6INO: C 34.03, H 2.45, N 5.67; found: C. 34.23; H. 2.22; N. 5.51.

2-Iodophenyl isocyanide (159-I)

To a solution of N-formyl-2-iodoaniline (5.11 g, 20.7 mmol) in anhydrous CH_2Cl_2 (130 mL) was added at 0 °C diisopropylamine (17 mL, 120 mmol), then dropwise over a period of 10 min $POCl_3$ (4.4 mL, 41.4 mmol). The mixture was stirred at 0 °C for 15 min, then a saturated solution of Na_2CO_3 (40 mL) was added slowly. The mixture was transferred into a separatory funnel, diluted with dichloromethane (200 mL), the organic phase washed with a half-saturated solution of Na_2CO_3 (100 mL) and brine (100 mL), then dried over anhydrous Na_2SO_4 and evaporated. The crude product was purified by recrystallization from hexane to give 4.43 g (93%) of the title compound as a colorless solid, m. p. 42 °C. R_f = 0.29 (hexane/EtOAc 30 : 1). ^{1}H NMR (300 MHz, CDCl$_3$): δ 7.90 (d, J = 7.8 Hz, 1 H), 7.44–7.36 (m, 2 H), 7.14–7.09 (m, 1 H); ^{13}C NMR (75.5 MHz, CDCl$_3$): δ 167.4 (C), 139.6 (CH), 130.4 (2 CH), 129.0 (CH), 127.6 (C), 109.7 (C); MS (70eV, EI) m/z (%): 229.0 (100) [M$^+$], 57.1(92), 71.1(80),

97.2(70); IR (KBr): ν = 2123 (NC), 1459, 1434, 1042,1019, 751, 643, 432 cm^{-1}; Anal. Calcd for C$_7$H$_4$IN: C 36.71, H 1.76, N 6.12; found: C 36.88, H 1.88, N 5.87.

N-Formyl-2-bromoaniline

The title compound was prepared in the same way as N-formyl-2-iodoaniline from 26.5 g (154 mmol) of 2-bromoanilin, 63 mL of ethyl formate and 7.7 g of a 60% suspension of NaH in mineral oil (193 mmol) to give, after washing with hexane, 28.6 (93%) of pure product as a colorless solid, m. p. 93 °C. R_f = 0.41 (hexane/EtOAc 2 : 1). ^{1}H NMR (300 MHz, DMSO-d6): δ 9.67 (br s, 1 H, NH), 8.36 (s, 1 H, CHO), 8.01 (d, J = 7.9 Hz, 1 H), 7.64 (d, J = 8.3 Hz, 1 H), 7.35 (t, J = 7.3 Hz, 1 H), 6.94 (t, J = 7.2 Hz, 1 H); ^{13}C NMR (75.5 MHz, DMSO-d6): δ 160.3 (CH), 135.4 (C), 132.6 (CH), 128.0 (CH), 126.0 (CH), 123.9 (CH), 114.4 (C); MS (70 eV, EI) m/z (%): 199.1 (18) [M$^+$], 171.0 (22), 120.1 (100), 92.1 (40), 65.1 (36); IR (KBr): 3256, 2904, 1666 (C=O), 1600, 1578, 1536, 1437, 1401, 1292, 1157, 1117, 1024, 861, 742, 654, 528, 433 cm^{-1}; Anal. Calcd for C$_7$H$_6$BrNO: C 42.03, H 3.02, N 7.00; found: C 42.35, H 2.88, N 6.80.

2-Bromophenyl isocyanide (159-Br)[186]

2-Bromophenyl isocyanide **159**-Br was prepared in the same way as 2-iodophenyl isocyanide from N-formyl-2-bromoaniline (28.6 g, 143 mmol), diisopropylamine (116 mL, 829 mmol) and POCl$_3$ (30.3 mL, 286 mmol) to give, after recrystallisation from hexane, 22.0 g (85%) of pure product as a colorless solid, m. p. 40 °C. [lit[186] 41 °C] R_f = 0.4 (hexane/EtOAc 10 : 1). ^{1}H NMR (300 MHz, CDCl$_3$): δ = 7.66 (dd, J = 8.3, 1.5 Hz, 1 H), 7.44 (dd, J = 7.9, 1.5 Hz, 1 H), 7.36 (dt, J = 7.9, 1.5 Hz, 1 H), 7.28 (dt, J = 7.9, 1.9 Hz, 1 H); ^{13}C NMR (75.5 MHz, CDCl$_3$): δ 168.3 (C), 133.1 (CH), 130.3 (2 CH), 128.1 (C), 128.0 (CH), 119.7 (C); MS (70eV, EI) m/z (%): 181.1 (18) [M$^+$], 120.1(58), 102.1(100), 91.1(36); IR (KBr): 2125 (NC), 1468, 1047, 753, 439 cm^{-1}; Anal. Calcd for C$_7$H$_4$BrN: C 46.19, H 2.22, N 7.70; found: C 46.38, H 2.10, N 7.61.

General Procedure for the Bromine-Lithium Exchange Reaction of 2-Bromophenyl Isocyanide 159-Br and Subsequent Trapping with Electrophiles to Give 2-Substituted Phenyl Isocyanides 192 (GP5)

To a solution of 2-bromophenyl isocyanide **159**-Br (364 mg, 2 mmol) in anhydrous tetrahydrofuran (20 mL), kept in an oven-dried 25 mL-Schlenk flask under an atmosphere of dry nitrogen, was added dropwise with stirring a 2.5 M solution of *n*BuLi in hexane (0.8 mL, 2 mmol) at −78 °C over a period of 5 min. The mixture was stirred at −78 °C for an additional 10 min, then the electrophile (2 mmol) in anhydrous THF (2 mL) was added dropwise. The mixture was stirred at −78 °C for 3 h, and the reaction was quenched with saturated NH_4Cl solution (2 mL). The mixture was warmed to r.t., diluted with diethyl ether (50 mL), the organic phase washed with water (2 × 10 mL), brine (20 mL) and dried over anhydrous Na_2SO_4. The solvents were removed under reduced pressure to give a crude product, which was purified by column chromatography on silica gel or Kugelrohr distillation.

General Procedure for the Synthesis of 3-Substituted Quinazolin-4(3*H*)-ones (-thiones) (GP6)

To a solution of 2-bromophenyl isocyanide **159**-Br (364 mg, 2 mmol) in anhydrous tetrahydrofuran (20 mL), kept in an oven-dried 25 mL-Schlenk flask under an atmosphere of dry nitrogen, was added dropwise with stirring a 2.5 M solution of *n*BuLi in hexane (0.8 mL, 2 mmol) at −78 °C over a period of 5 min. The mixture was stirred at −78 °C for an additional 10 min, and then the respective isocyanate (2 mmol) in anhydrous THF (2 mL) was added dropwise. The mixture was stirred at −78 °C for 3h and the reaction quenched with saturated NH_4Cl solution (2 mL). The mixture was warmed to r.t., diluted with diethyl ether (50 mL), the organic phase washed with water (2 × 10 mL), brine (20 mL) and dried over anhydrous Na_2SO_4. The solvents were removed under reduced pressure to give a crude product, which was purified by column chromatography on silica gel.

General Procedure for the Synthesis of 2,3-Disubstituted Quinazolin-4(3*H*)-ones (GP7)

The procedure is the same as GP5, but the intermediate was trapped by addition at −78 °C of the respective second electrophile (2 mmol), and after stirring at −78 °C for an additional 1 h, the reaction mixture was warmed gradually to r.t., diluted with diethyl ether (50 mL), washed with water (2 × 10 mL), brine (20 mL) and dried over anhydrous Na_2SO_4. The solvents were removed under reduced pressure to give a crude product, which was purified by column chromatography on silica gel.

2-Iodophenyl isocyanide (159-I)

The isocyanide **159**-I (403 mg, 88%) was obtained from *o*-bromophenyl isocyanide (**159**-Br) (364 mg, 2 mmol) and iodine (708 mg, 2 mmol) following GP5, after column chromatography (hexane/EtOAc 30 : 1, R_f = 0.29). The analytical data are identical to those of an authentic sample described above.

Methyl 2-isocyanobenzoate (192a)[187]

The isocyanide **192a** (254 mg, 79%) was obtained from 2-bromophenyl isocyanide (**159**-Br) (364 mg, 2 mmol) and methyl chloroformate (189 mg, 2 mmol) following GP5, after column chromatography (hexane/EtOAc 5 : 1, R_f = 0.29) as a yellow oil, which turned dark upon standing at r.t. ^{1}H NMR (300 MHz, CDCl$_3$): δ 8.01 (d, *J* = 7.2 Hz, 1 H, Ar-CH), 7.61–7.45 (m, 3 H, Ar-CH), 3.99 ppm (s, 3 H, CH$_3$); ^{13}C NMR (75.5 MHz, CDCl$_3$): δ 169.4 (C), 164.5 (C), 133.0 (CH), 131.3 (CH), 129.2 (CH), 128.9 (CH), 128.6 (C), 127.1 (C), 52.7 ppm (CH$_3$); MS (70 eV, EI) *m/z* (%): 161.2 (44) [M$^+$], 146.2 (42), 130.2 (100), 102.2 (74); IR (KBr): 2954, 2126 (NC), 1734 (C=O), 1598, 1488, 1436, 1268, 1135, 1082, 759 cm^{-1}; HRMS (ESI) calcd for C$_9$H$_7$NO$_2^+$ [M+H$^+$]: 162.05495; found: 162.05501.

2-(Phenylthio)phenyl isocyanide (192b)

The isocyanide **192b** (354 mg, 84%) was obtained from 2-bromphenyl isocyanide (**159**-Br) (364 mg, 2 mmol) and diphenyl disulfide (436 mg, 2 mmol) following GP5, after Kugelrohr distillation (120–130 °C, 0.3 Torr) as a yellow oil. ^{1}H NMR (300 MHz, CDCl$_3$): δ 7.50–7.37 (m, 6 H, Ar-H), 7.25–7.17 (m, 2 H, Ar-H), 7.04 ppm (dd, *J* = 7.2, 2.2 Hz, 1 H, Ar-H); ^{13}C NMR (75.5 MHz, CDCl$_3$): δ 168.4 (C), 135.3 (C), 133.6 (2 CH), 131.4 (CH), 129.7 (2 CH), 129.6 (CH), 129.0 (C), 128.8 (CH), 127.4 (C), 127.3 (CH), 126.8 ppm (CH); MS (70 eV, EI) *m/z* (%): 211.2 (100) [M$^+$], 184.2 (22); IR (KBr): 3061, 2117 (NC), 1581, 1464, 1440, 1066, 1024, 751, 690 cm^{-1}; elemental analysis calcd (%) for C$_{13}$H$_9$NS: C 73.90, H 4.29, N 6.63; found: C 73.98, H 4.21, N 6.61

2-Isocyanobenzaldehyde (192c)

The isocyanide **192c** (145 mg, 55%) was obtained from 2-bromphenyl isocyanide (**159**-Br) (364 mg, 2 mmol) and methyl formate (120 mg, 2 mmol) following GP5, after column chromatography (hexane/EtOAc 10 : 1, R_f = 0.15) as a colorless solid, m. p. 49–50 °C. ^{1}H NMR (300 MHz, CDCl$_3$): δ 10.44 (s, 1 H, CHO), 7.97 (d, J = 7.8 Hz, 1 H, Ar-H), 7.69 (t, J = 7.8 Hz, 1 H, Ar-H), 7.59 (d, J = 7.8 Hz, 1 H, Ar-H), 7.54 ppm (t, J = 7.8 Hz, 1 H, Ar-H); ^{13}C NMR (75.5 MHz, CDCl$_3$, APT): δ 187.7 (CH), 170.7 (C), 135.3 (C), 134.9 (CH), 129.9 (C), 129.9 (CH), 128.7 (CH), 127.9 ppm (CH); MS (DCI) m/z (%): 132.0 (100) [M+H$^+$]; IR (KBr): 2894, 2119 (NC), 1705 (C=O), 1594, 1476, 1455, 1411, 1403, 1272, 1201, 1090, 831, 637 cm^{-1}; HRMS (ESI) calcd for C$_8$H$_5$NO$^+$ [M+H$^+$]: 132.04439; found: 132.04443.

2-(Formylamino)benzaldehyde (196)

The compound **196** (225 mg, 76%) was obtained from 2-bromophenyl isocyanide (**159**-Br) (364 mg, 2 mmol) and dimethyl formamide (146 mg, 2 mmol) following GP5 and after column chromatography (hexane/EtOAc 2 : 1, R_f = 0.30) as a colorless solid, m. p. 74–75 °C [lit.[188] 75–77] ^{1}H NMR (300 MHz, CDCl$_3$): δ 11.05 (br s, 1 H, NH), 9.94 (s, 1 H, CHO), 8.73 (d, J = 8.7 Hz, 1 H, Ar-H), 8.54 (s, 1 H, NCHO), 7.71 (d, J = 7.5 Hz, 1 H, Ar-H), 7.63 (t, J = 7.2 Hz, 1 H, Ar-H), 7.29 ppm (t, J = 7.5 Hz, 1 H, Ar-H); ^{13}C NMR (75.5 MHz, CDCl$_3$, APT): δ 195.4 (CH), 159.9 (CH), 139.6 (C), 136.2 (CH), 136.0 (CH), 123.6 (CH), 121.7 (C), 120.7 ppm (CH); MS (70 eV, EI) m/z (%): 149.2 (50) [M$^+$], 121.2 (68), 93.2 (100); IR (KBr): 3274, 1671 (C=O), 1596, 1528, 1456, 1406, 1291, 1195, 1166, 1147, 875, 757 cm^{-1}; elemental analysis calcd (%) for C$_8$H$_7$NO$_2$: C 64.42, H 4.73, N 9.39; found: C 64.35, H 4.60, N 9.52.

(2-Isocyanophenyl) (2-carbomethoxyphenyl) ketone (192d)

The isocyanide **192d** (419 mg, 79%) was obtained from 2-bromophenyl isocyanide (**159**-Br) (364 mg, 2 mmol) and methyl phthaloyl chloride (397 mg, 2 mmol) following GP5, after column chromatography (hexane/EtOAc 2 : 1, R_f = 0.21) as a yellow solid, m. p. 76–77 °C. ^{1}H NMR (300 MHz, CDCl$_3$): δ 8.00 (dd, J = 7.5, 1.5 Hz, 1 H, Ar-CH), 7.70–7.40 (m, 7 H, Ar-CH), 3.69 ppm (s, 3 H, CH$_3$); ^{13}C NMR (75.5 MHz, CDCl$_3$, APT): δ 193.7 (C), 169.9 (C), 166.5 (C), 140.9 (C), 133.9 (C), 132.8 (CH), 132.4 (CH), 130.7 (2 CH), 130.2 (CH),

129.7 (2 C), 129.1 (CH), 129.0 (CH), 128.1 (CH), 52.4 ppm (CH$_3$); MS (DCI) m/z (%): 283.3 (8) [M+NH$_4^+$], 266.3 (100) [M+H$^+$]; IR (KBr): 2125 (NC), 1718 (C=O), 1684 (C=O), 1595, 1578, 1284, 1083, 933, 756, 717 cm^{-1}; HRMS (ESI) calcd for C$_{16}$H$_{12}$NO$_3^+$ [M+H$^+$]: 266.08117; found: 266.08121.

3-Phenylquinazolin-4(3H)-one (191a)

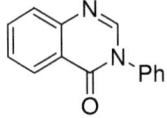

The quinazolin-4(3H)-one **191a** (404 mg, 91%) was obtained from 2-bromophenyl isocyanide (**159-Br**) (364 mg, 2 mmol) and phenyl isocyanate (238 mg, 2 mmol) following GP6, after column chromatography (hexane/EtOAc 2 : 1, R$_f$ = 0.15) as a colorless solid, m. p. 135–136 °C. [lit.[189] 137–137.5 °C] ^{1}H NMR (300 MHz, CDCl$_3$): δ 8.37 (d, J = 8.1 Hz, 1 H), 8.13 (s, 1 H), 7.84–7.75 (m, 2 H), 7.59–7.49 (m, 4 H), 7.45–7.41 (m, 2 H); ^{13}C NMR (75.5 MHz, CDCl$_3$, APT): δ 160.7 (C), 147.8 (C), 146.0 (CH), 137.4 (C), 134.5 (2 CH), 129.6 (2 CH), 129.1 (CH), 127.7 (CH), 127.6 (CH), 127.1 (CH), 127.0 (CH), 122.3 ppm (C); MS (70 eV, EI) m/z (%): 222.3 (100) [M$^+$]; IR (KBr): 3067, 3048, 2360, 2338, 1618 (C=O), 1227, 1128, 700 cm^{-1}; elemental analysis calcd (%) for C$_{14}$H$_{10}$N$_2$O: C 75.66, H 4.54, N 12.60; found: C 75.45, H 4.60, N 12.38.

3-p-Tolylquinazolin-4(3H)-one (191b)

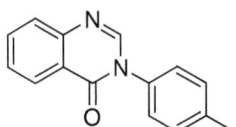

The compound **191b** (420 mg, 89%) was obtained from 2-bromophenyl isocyanide (**159-Br**) (364 mg, 2 mmol) and p-tolyl isocyanate (266 mg, 2 mmol) following GP6, after column chromatography (hexane/EtOAc 1 : 1, R$_f$ = 0.33) as a colorless solid, m. p. 143–144 °C. [lit.[190] 147 °C] ^{1}H NMR (300 MHz, CDCl$_3$): δ 8.37 (d, J = 8.7 Hz, 1 H), 8.12 (s, 1 H), 7.84–7.75 (m, 2 H), 7.55 (dd, J = 6.4, 1.9 Hz, 1 H), 7.37–7.29 (m, 5 H); ^{13}C NMR (75.5 MHz, CDCl$_3$, APT): δ 160.9 (C), 148.0 (C), 146.3 (CH), 139.2 (C), 135.0 (C), 134.5 (CH), 130.2 (2 CH), 127.6 (CH), 127.5 (CH), 127.2 (CH), 126.8 (2 CH), 122.5 (C), 21.2 ppm (CH$_3$); MS (70 eV, EI) m/z (%): 236.0 (100) [M$^+$]; IR (KBr): 1688 (C=O), 1600, 1514, 1471, 1322, 1292, 1260, 1193, 917, 817, 770, 750, 694, 616, 556, 521, 482 cm^{-1}; elemental analysis calcd (%) for C$_{15}$H$_{12}$N$_2$O: C 76.25, H 5.12, N 11.86; found: C 75.96, H 4.96, N 12.11.

3-(4-(Trifluoromethyl)phenyl)quinazolin-4(3*H*)-one (191c)

The compound **191c** (400 mg, 69%) was obtained from 2-bromophenyl isocyanide (**159**-Br) (364 mg, 2 mmol) and 4-(trifluoromethyl)phenyl isocyanate (374 mg, 2 mmol) following GP6, after column chromatography (hexane/EtOAc 2 : 1, R_f = 0.21) as a colorless solid, m. p. 183–184 °C. [lit.[190] 132 °C] ^{1}H NMR (300 MHz, CDCl$_3$): δ 8.39 (d, *J* = 7.5 Hz, 1 H, Ar-H), 8.12 (s, 1 H, CH=N), 7.87–7.78 (m, 4 H, Ar-H), 7.60 (m, 3 H, Ar-H); ^{13}C NMR (75.5 MHz, CDCl$_3$, APT): δ 160.5 (C), 147.7 (C), 145.1 (CH), 140.4 (C), 135.0 (2 CH), 128.0 (CH), 127.8 (CH), 127.5 (2 CH), 127.2 (CH), 126.9 (q, J_{CF} = 3.9 Hz, C), 126.3 (CH), 122.1 ppm (C); MS (70 eV, EI) *m/z* (%): 290.2 (100) [M$^+$], 145.0 (16), 119.0 (15); IR (KBr): 3440 (br), 1615 (C=O), 1325, 1167, 1113, 1064 cm^{-1}; elemental analysis calcd (%) for C$_{15}$H$_9$F$_3$N$_2$O: C 62.07, H 3.13, N 9.65; found: C 61.96, H 3.11, N 10.01.

3-(4-Fluorophenyl)quinazolin-4(3*H*)-one (191d)

The compound **191d** (360 mg, 75%) was obtained from 2-bromophenyl isocyanide (**159**-Br) (364 mg, 2 mmol) and 4-fluorophenyl isocyanate (274 mg, 2 mmol) following GP6, after column chromatography (hexane/EtOAc 2 : 1, R_f = 0.21) as a colorless solid, m. p. 189–190 °C. [lit.[191] 170–171 °C] ^{1}H NMR (300 MHz, CDCl$_3$): δ 8.36 (dd, *J* = 8.3, 1.1 Hz, 1 H), 8.10 (s, 1 H), 7.85–7.76 (m, 2 H), 7.59–7.54 (m, 2 H), 7.44–7.39 (m, 2 H), 7.28–7.20 ppm (m, 2 H); ^{13}C NMR (75.5 MHz, CDCl$_3$, APT): δ 161.0 (C), 147.8 (C), 145.8 (CH), 134.7 (CH), 133.4 (C), 129.0 (CH), 128.8 (CH), 127.8 (CH), 127.7 (CH), 127.2 (CH), 122.2 (C), 116.8 (CH), 116.5 ppm (CH); MS (70 eV, EI) *m/z* (%): 240.2 (100) [M$^+$], 212.1 (10), 119.1 (20), 95.1 (18); IR (KBr): 1660 (C=O), 1614, 1511, 1469, 1405, 1329, 1293, 1262, 1227, 1102, 927, 833, 775, 697, 613, 552, 526, 436 cm^{-1}; elemental analysis calcd (%) for C$_{14}$H$_9$FN$_2$O: C 69.99, H 3.78, N 11.66; found: 69.79, H 4.03, N 11.80.

3-Benzylquinazolin-4(3*H*)-one (191e)

The compound **191e** (350 mg, 74%) was obtained from 2-bromophenyl isocyanide (**159**-Br) (364 mg, 2 mmol) and benzyl isocyanate (266 mg, 2 mmol) following GP6, after column chromatography (hexane/EtOAc 1 : 1, R_f = 0.25) as a colorless solid, m. p. 116–117 °C [lit.[192] 117–118 °C]. ^{1}H NMR (300 MHz, CDCl$_3$): δ 8.33 (d, *J* = 7.5 Hz, 1 H), 8.11 (s, 1 H), 7.78–7.69 (m, 2 H),

7.51 (t, J = 8.0 Hz, 1 H), 7.35 (m, 5 H); ^{13}C NMR (75.5 MHz, CDCl$_3$, APT): δ 161.1 (C), 148.1 (C), 146.3 (CH), 135.8 (C), 134.2 (CH), 129.0 (2 CH), 128.3 (CH), 128.0 (2 CH), 127.6 (CH), 127.3 (CH), 126.9 (CH), 122.3 (C), 49.6 ppm (CH$_2$); MS (70 eV, EI) m/z (%): 236.0 (100) [M$^+$], 130.1 (27), 91.0 (70); IR (KBr): 1684 (C=O), 1605, 1475, 1365, 1321, 1150, 938, 774, 747, 706, 694, 606 cm^{-1}; elemental analysis calcd (%) for C$_{15}$H$_{12}$N$_2$O: C 76.25, H 5.12, N 11.86; found: C 76.19, H 5.28, N 12.10.

3-Isopropylquinazolin-4(3*H*)-one (191f)

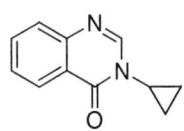

The compound **191f** (306 mg, 81%) was obtained from 2-bromophenyl isocyanide (**159**-Br) (364 mg, 2 mmol) and isopropyl isocyanate (170 mg, 2 mmol) following GP6, after column chromatography (hexane/EtOAc 2 : 1, R_f = 0.10) as a yellow solid, m. p. 87–88 °C [lit[193] 88–89 °C]. ^{1}H NMR (300 MHz, CDCl$_3$): δ 8.32 (d, J = 7.9 Hz, 1 H, Ar-CH), 8.13 (s, 1 H, CH=N), 7.78–7.69 (m, 2 H, Ar-CH), 7.50 (dd, J = 6.4, 1.5 Hz, 1 H, Ar-CH), 5.21 (m, 1 H, CH(CH$_3$)$_2$), 1.50 (d, J = 6.8 Hz, 6 H, CH$_3$); ^{13}C NMR (75.5 MHz, CDCl$_3$, APT): δ 160.6 (C), 147.5 (C), 143.5 (CH), 134.1 (CH), 127.2 (CH), 127.1 (CH), 126.8 (CH), 121.9 (C), 45.9 (CH$_3$), 22.0 ppm (CH$_3$); MS (70 eV, EI) m/z (%): 188.1 (44) [M$^+$], 146.0 (100), 117.9 (14); IR (KBr): 3415 (br), 1640 (C=O), 1180, 1130, 1076, 773 cm^{-1}; elemental analysis calcd (%) for C$_{11}$H$_{12}$N$_2$O: C 70.19, H 6.43, N 14.88; found: C 69.96, H 6.39, N 14.49.

3-Cyclopropylquinazolin-4(3*H*)-one (191g)

The compound **191g** (260 mg, 70%) was obtained from 2-bromophenyl isocyanide (**159**-Br) (364 mg, 2 mmol) and cyclopropyl isocyanate (166 mg, 2 mmol) following GP6, after column chromatography (hexane/EtOAc 1 : 1, R_f = 0.18) as a colorless solid, m. p. 96–97 °C. ^{1}H NMR (300 MHz, CDCl$_3$): δ 8.31 (d, J = 7.9 Hz, 1 H, Ar-H), 8.11 (s, 1 H, CH=N), 7.78–7.67 (m, 2 H, Ar-H), 7.50 (t, J = 6.8 Hz, 1 H, Ar-H), 3.29–3.22 (m, 1 H, cPr-CH), 1.25–1.18 (m, 2 H, cPr-CH$_2$), 0.97–0.91 (m, 2 H, cPr-CH$_2$); ^{13}C NMR (75.5 MHz, CDCl$_3$, APT): δ 162.2 (C), 147.6 (C), 146.7 (CH), 134.1 (CH), 127.3 (CH), 127.2 (CH), 126.6 (CH), 121.9 (C), 29.2 (CH), 6.4 ppm (CH$_2$); MS (70 eV, EI) m/z (%): 186.1 (100) [M$^+$], 171.0 (48); IR (KBr): 3424 (br), 1648 (C=O), 1561, 1470, 1259, 1176, 1105, 773 cm^{-1}; elemental analysis calcd (%) for C$_{11}$H$_{10}$N$_2$O: C 70.95, H 5.41, N 15.04; found: C 70.73, H 5.70, N 14.86.

3-Cyclopropylquinazoline-4(3*H*)-thione (191h)

The compound **191h** (287 mg, 71%) was obtained from 2-bromophenyl isocyanide (**159**-Br) (364 mg, 2 mmol) and isopropyl isothiocyanate (198 mg, 2 mmol) following a modified GP6 (the reaction was quenched by addition of water at −40 °C), after column chromatography (hexane/EtOAc 5 : 1, R_f = 0.38) as a yellow solid, m. p. 60–61 °C. ^{1}H NMR (300 MHz, CDCl$_3$): δ 8.61 (s, 1 H, CH=N), 8.21 (d, *J* = 8.3 Hz, 1 H), 7.57–7.51 (m, 2 H), 7.38 (td, *J* = 8.3, 6.0, 2.6 Hz, 1 H, Ar-H), 2.93–2.86 (m, 1 H), 1.05–0.97 (m, 2 H, cPr-CH$_2$), 0.94–0.89 (m, 2 H, cPr-CH$_2$); ^{13}C NMR (75.5 MHz, CDCl$_3$, APT): δ 148.9 (CH), 147.7 (C), 143.0 (C), 131.7 (CH), 130.4 (CH), 129.1 (CH), 124.4 (CH), 121.8 (C), 35.4 (CH), 8.7 ppm (CH$_2$); MS (70 eV, EI) *m/z* (%): 202.1 (42) [M$^+$], 187.1 (56), 174.0 (72), 169.1 (56), 147.0 (49), 120.0 (100); IR (KBr): 1584 (C=S), 1550, 1469, 1445, 1266, 1158, 1020, 937, 856, 765, 750 cm^{-1}; elemental analysis calcd (%) for C$_{11}$H$_{10}$N$_2$S: C 65.32, H 4.98, N 13.85; found: C 65.13, H 4.80, N 13.55.

3-Cyclohexylquinazoline-4(3*H*)-thione (191i)

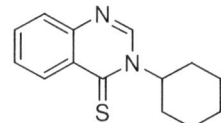

The compound **191i** (380 mg, 78%) was obtained from 2-bromophenyl isocyanide (**159**-Br) (364 mg, 2 mmol) and cyclohexyl isothiocyanate (282 mg, 2 mmol) following a modified GP6 (the reaction was quenched by addition of water at −40 °C), after column chromatography (hexane/EtOAc 5 : 1, R_f = 0.42) as a yellow solid, m. p. 91–92 °C. ^{1}H NMR (300 MHz, CDCl$_3$): δ 8.61 (s, 1 H, CH=N), 8.31 (d, *J* = 7.9 Hz, 1 H, Ar-H), 7.56 (d, *J* = 3.4 Hz, 2 H, Ar-H), 7.44–7.37 (m, 1 H, Ar-H), 3.45–3.36 (m, 1 H, CH), 1.87–1.26 (m, 10 H, CH$_2$); ^{13}C NMR (75.5 MHz, CDCl$_3$, APT): δ 149.4 (CH), 144.9 (C), 143.5 (C), 131.7 (CH), 130.4 (CH), 129.0 (CH), 125.1 (CH), 122.0 (C), 61.6 (CH), 32.3 (2 CH$_2$), 25.8 (CH$_2$), 22.0 ppm (2 CH$_2$); MS (70 eV, EI) *m/z* (%): 244.1 (80) [M$^+$], 211.1 (70), 162.0 (80), 129.1 (100); IR (KBr): 2927, 2854, 1591 (C=S), 1554, 1444, 1362, 1268, 1068, 959, 936, 846, 763, 606 cm^{-1}; elemental analysis calcd (%) for C$_{14}$H$_{16}$N$_2$S: C 68.81, H 6.60, N 11.46; found: C 68.75, H 6.40, N 11.20.

Methyl 3,4-dihydro-4-oxo-3-phenylquinazoline-2-carboxylate (191j)

The compound **191j** (412 mg, 73%) was obtained from 2-bromophenyl isocyanide (**159**-Br) (364 mg, 2 mmol), phenyl isocyanate (238 mg, 2 mmol) and methyl chloroformate (208 mg, 2 mmol) following GP7, after column chromatography (hexane/EtOAc 2 : 1, R_f = 0.32) as a colorless solid. ^{1}H NMR (300 MHz, CDCl$_3$): δ 7.57–7.36 (m, 9 H, Ar-H), 3.68 ppm (s, 3 H, CH$_3$); ^{13}C NMR (75.5 MHz, CDCl$_3$, APT): δ 168.5 (C), 168.2 (C), 154.2 (C), 137.0 (C), 134.6 (C), 130.8 (CH), 129.6 (CH), 129.4 (C), 128.8 (CH), 128.3 (CH), 128.1 (CH), 126.7 (CH), 121.7 (C), 54.3 ppm (CH$_3$); MS (EI) m/z (%): 280.2 (37) [M$^+$], 130.2 (100), 119.1 (32), 102.1 (38); IR (KBr): 2126, 1753 (C=O), 1693 (C=O), 1597, 1493, 1433, 1322, 1260, 1054, 771, 749, 693, 632 cm^{-1}; HRMS (ESI) calcd for C$_{16}$H$_{12}$N$_2$O$_3$Na$^+$ [M+Na$^+$]: 303.07401; found: 303.07403.

3-Phenyl-2-(phenylthio)quinazolin-4(3H)-one (191k)

The compound **191k** (505 mg, 77%) was obtained from 2-bromophenyl isocyanide (**159**-Br) (364 mg, 2 mmol), phenyl isocyanate (238 mg, 2 mmol) and diphenyl disulfide (436 mg, 2 mmol) following GP7, after column chromatography (hexane/EtOAc 5 : 1, R_f = 0.26) as a colorless solid, m. p. 130–131 °C. ^{1}H NMR (300 MHz, CDCl$_3$): δ 8.21 (dd, J = 7.8, 1.6 Hz, 1 H), 7.64–7.52 (m, 6 H), 7.45–7.32 ppm (m, 7 H); ^{13}C NMR (75.5 MHz, CDCl$_3$, APT): δ 161.9 (C), 157.1 (C), 147.7 (C), 136.0 (C), 135.7 (2 CH), 134.4 (CH), 130.0 (CH), 129.7 (2 CH), 129.5 (CH), 129.2 (2 CH), 129.0 (2 CH), 128.5 (C), 127.1 (CH), 126.6 (CH), 126.0 (CH), 119.9 ppm (C); MS (EI) m/z (%): 330.3 (100) [M$^+$], 221.2 (36), 77.1 (28), 44.1 (28); IR (KBr): 1696 (C=O), 1540, 1467, 1296, 1260, 959, 764 cm^{-1}; HRMS (ESI) calcd for C$_{20}$H$_{15}$N$_2$OS$^+$ [M+H$^+$]: 331.08996; found: 331.09010.

2-Cyano-3-(phenyl)quinazolin-4(3H)-one (191l)

The compound **191l** (123 mg, 54%) was obtained from 2-bromophenyl isocyanide (**159**-Br) (182 mg, 1 mmol), phenyl isocyanate (119 mg, 1 mmol) and p-toluenesulfonyl cyanide (181 mg, 1 mmol) following GP7, after column chromatography (hexane/EtOAc 4 : 1, R_f = 0.16) as a colorless solid, m. p. 195–196 °C.[lit.[194] 198 °C] ^{1}H NMR (300 MHz, CDCl$_3$): δ 7.69–7.45 ppm (m, 9 H); ^{13}C NMR (75.5 MHz, CDCl$_3$, APT): δ 170.7 (C), 165.0 (C), 134.2 (C), 132.9 (CH), 129.9 (2 CH), 129.7 (CH), 129.6 (CH), 129.4 (C), 128.9 (C), 128.7 (CH), 127.7 (CH), 125.5 (2 CH), 120.0 ppm (C); MS (EI) m/z (%): 247.3 (32) [M$^+$], 130.2 (100), 102.2 (42); IR (KBr): 2239 (CN), 2131, 1734 (C=O),

1593, 1484, 1270, 1162, 1053, 754, 690 cm^{-1}; elemental analysis calcd (%) for $C_{15}H_9N_3O$: C 72.87, H 3.67, N 16.99; found: 72.79, H 3.66, N 16.75.

3-Benzyl-2-iodoquinazolin-4(3*H*)-one (191m)

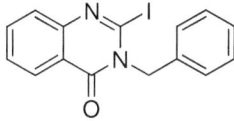

The compound **191m** (543 mg, 75%) was obtained from 2-bromophenyl isocyanide (**159**-Br) (364 mg, 2 mmol), benzyl isocyanate (266 mg, 2 mmol) and iodine (508 mg, 2 mmol) following GP7, after column chromatography (hexane/EtOAc 2 : 1, R_f = 0.39) as a colorless solid, m. p. 151–152 °C. ^{1}H NMR (300 MHz, CDCl$_3$): δ 8.27 (dd, *J* = 8.9, 1.5 Hz, 1 H), 7.75 (dt, *J* = 7.8, 1.5 Hz, 1 H), 7.65 (d, *J* = 6.8 Hz 1 H), 7.51 (dt, *J* = 7.6, 1.1 Hz, 1 H), 7.35–7.29 (m, 5 H), 5.56 ppm (s, 2 H, CH$_2$); ^{13}C NMR (75.5 MHz, CDCl$_3$): δ 160.4 (C), 135.3 (C), 134.8 (CH), 128.7 (CH), 127.8 (2 CH), 127.3 (CH), 127.2 (CH), 126.9 (CH), 121.0 (C), 112.9 (C), 55.6 ppm (CH$_2$); MS (EI) *m/z* (%): 362.2 (100) [M$^+$], 235.2 (52), 91.1 (52); IR (KBr): 1671 (C=O), 1542, 1467, 1331, 1150, 1075, 957, 771 cm^{-1}; elemental analysis calcd (%) for $C_{15}H_8N_2O_2$: C. 49.75, H. 3.06, N 7.73; found: 49.44, H 3.25, N 7.99.

2,3-Dihydropyrrolo[2,1-b]quinazolin-9(1*H*)-one (desoxyvascinone, 191n)

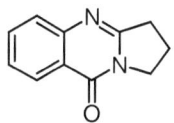

The compound **191n** (268 mg, 72%) was obtained from 2-bromophenyl isocyanide (**159**-Br) (364 mg, 2 mmol) and 3-iodopropyl isocyanate (422 mg, 2 mmol) following GP7, after column chromatography (EtOAc, R_f = 0.11) as a colorless solid, m. p. 104–105 °C [lit.$^{[195]}$ 105–107 °C]. ^{1}H NMR (300 MHz, CDCl$_3$): δ 8.27 (dd, *J* = 8.4, 1.6 Hz, 1 H, Ar-H), 7.72 (t, *J* = 8.4 Hz, 1 H, Ar-H), 7.63 (d, *J* = 8.1 Hz 1 H, Ar-H), 7.44 (t, *J* = 8.1 Hz, 1 H), 4.20 (dd, *J* = 7.5, 7.5 Hz, 2 H), 3.17 (dd, *J* = 7.8, 7.8 Hz, 2 H), 2.34–2.23 (m, 2 H) ppm; ^{13}C NMR (75.5 MHz, CDCl$_3$, APT): δ 160.9 (C), 159.3 (C), 149.0 (C), 134.0 (CH), 126.7 (CH), 126.2 (CH), 126.1 (CH), 120.3 (C), 46.4 (CH$_2$), 32.4 (CH$_2$), 19.4 ppm (CH$_2$); MS (DCI) *m/z* (%): 373.4 (40) [2M+H$^+$], 204.2 (70) [M+NH$_4^+$], 187.2 (100) [M+H$^+$]; IR (KBr): 2924, 1675 (C=O), 1621, 1465, 1384, 1336, 1268, 1022, 771, 694 cm^{-1}; elemental analysis calcd (%) for $C_{11}H_{10}N_2O$: C 70.95, H 5.41, N 15.04; found: 71.13, H 5.09, N 14.80.

Indolo[2,1-b]quinazoline-6,12-dione (trypthamine, 191o)

The compound **191o** (210 mg, 85%) was obtained from 2-bromophenyl isocyanide (**159**-Br) (182 mg, 1 mmol) and methyl (2-isocyanato)benzoate (177 mg, 1 mmol) following GP7, after column chromatography (EtOAc, R_f = 0.55) as a yellow solid, m. p. 261–262 °C [lit.[196] 267–268 °C]. ^{1}H NMR (300 MHz, DMSO[d6], 100 °C): δ 8.46 (d, J = 8.3 Hz, 1 H, Ar-H), 8.29 (d, J = 7.9 Hz, 1 H, Ar-H), 7.90 (d, J = 3.8 Hz, 2 H, Ar-H), 7.83 (t, J = 7.2 Hz, 2 H, Ar-H), 7.73–7.68 (m, 1 H, Ar-H), 7.46 ppm (t, J = 7.5 Hz, 1 H); ^{13}C NMR (75.5 MHz, DMSO[d6], 100 °C): δ 181.6 (C), 157.1 (C), 146.0 (C), 145.5 (C), 144.3 (C), 137.2 (CH), 134.4 (CH), 129.4 (CH), 129.3 (CH), 126.4 (CH), 126.3 (CH), 124.1 (CH), 122.9 (C), 121.6 (C), 116.5 ppm (CH); MS (EI) m/z (%): 248.2 (100) [M$^+$], 220.2 (15); IR (KBr): 1725 (C=O), 1685 (C=O), 1594, 1458, 1353, 1312, 1190, 1116, 1039, 925, 755, 690cm^{-1}; elemental analysis calcd (%) for $C_{15}H_8N_2O_2$: C 72.58, H 3.25, N 11.28; found: 72.29, H 3.13, N 10.97.

Experimental Procedures for the Compounds Described in Chapter 3 "Reactions of ortho-Lithiophenyl (-Hetaryl) Isocyanides with Carbonyl Compounds – Rearrangements of 2-Metallated 4H-3,1-Benzoxazines"

N-(2-Bromopyridin-3-yl)formamide

To a solution of 3-amino-2-bromopyridine (1.31 g, 7.57 mmol) and ethyl formate (3 mL) in anhydrous THF (50 mL) was added portionwise at r.t. a suspension of NaH (60% in mineral oil, 378 mg, 9.46 mmol). The resulting mixture was stirred at r.t. for 24 h, and then the reaction was quenched with cold water (1 mL). The solvents were removed under reduced pressure, and the residue was dissolved in ethyl acetate/water (40/10 mL). The aqueous phase was extracted with ethyl acetate (2 × 30 mL), the combined organic extracts were dried over anhydrous Na_2SO_4 and concentrated. The residue was washed thoroughly with hexane (3 × 20 mL) and dried in vacuo to give 1.38 g (91%) of the title compound as a colorless solid, m.p. 139–140 °C. R_f = 0.13 (hexane/EtOAc 2 : 1). ^{1}H NMR (300 MHz, DMSO[d6]): δ 9.87 (br s, 1 H, NH), 8.41 (s, 1 H, CHO), 8.38 (d, J = 9.0 Hz, 1 H), 8.16 (d, J = 4.1 Hz, 1 H), 7.45 ppm (dd, J = 8.1, 4.7 Hz, 1 H); ^{13}C NMR (75.5 MHz, DMSO[d6]): δ 160.8 (CH), 145.4 (CH), 133.9 (C), 133.1 (C), 131.2 (CH), 123.6 ppm (CH); MS (70 eV, EI) *m/z* (%): 201.0 (2) [M$^+$], 200.0 (20), 121.1 (100), 93.1 (50); IR (KBr): 3243 (br), 1664 (C=O), 1585, 1517, 1449, 1400, 1379, 1288, 1150, 1117, 1047, 802, 735, 65 cm^{-1}; Anal. Calcd for $C_6H_5BrN_2O$: C 35.85, H 2.51, N 13.94; found: C 35.68, H 2.70, N 13.81.

2-Bromo-3-isocyanopyridine

To a solution of *N*-(2-bromopyridin-3-yl)formamide (1.35 g, 6.72 mmol) in anhydrous CH_2Cl_2 (45 mL) was added at 0 °C triethylamine (5.95 mL, 43 mmol), then dropwise over a period of 10 min $POCl_3$ (1.29 mL, 13.44 mmol). The mixture was stirred at 0 °C for 15 min, then a saturated solution of Na_2CO_3 (10 mL) was added slowly. The mixture was transferred into a separatory funnel, diluted with dichloromethane (50 mL), the organic phase washed with a half-saturated solution of Na_2CO_3 (100 mL) and brine (100 mL), then dried over anhydrous Na_2SO_4 and concentrated. The crude product was purified by column chromatography on silica gel (hexane/ethyl acetate 2 : 1, R_f = 0.30) to give 1.01 g (82%) of 2-bromo-3-isocyanopyridine as a colorless solid, m.p. 98–99 °C. ^{1}H NMR (300 MHz, CDCl$_3$): δ 8.43 (dd, J = 4.9, 1.9 Hz, 1 H), 7.76 (dd, J = 7.9, 1.9 Hz, 1 H), 7.39 ppm (dd, J = 7.9, 4.5 Hz, 1 H); ^{13}C NMR (75.5 MHz, CDCl$_3$): δ 172.8 (C), 149.6 (2 C), 139.3 (C), 135.6 (CH), 122.9 ppm (CH); MS

(70eV, EI) m/z (%): 185.0 (6) [M$^+$], 183.0 (12) [M+2$^+$], 110.1 (100), 105.1(50); IR (KBr): 2134 (NC), 1555, 1408, 1199,1064, 805, 722, 654, 517 cm^{-1}; Anal. Calcd for $C_6H_3BrN_2$: C 39.38, H 1.65, N 15.31; found: C 39.30, H 1.71, N 15.02.

N-(Thiophen-3-yl)formamide

To a stirred solution of 3-aminothiophene[178] (4.65 g, 47 mmol) and ethyl formiate (10 mL) in anhydrous THF (200 mL) was added portionwise at r.t. a suspension of NaH (60% in mineral oil, 2.26 g, 56.4 mmol). The resulting mixture was stirred at r.t. for 24 h, then the reaction was quenched with cold water (10 mL). The solvents were removed under reduced pressure, and the residue was dissolved in ethyl acetate/water (200/50 mL). The aqueous phase was extracted with ethyl acetate (2 × 50 mL), the combined organic extracts were dried over anhydrous Na_2SO_4 and concentrated to give almost pure product (5.80 g, 97%), which was used in the next step without further purification. R_f = 0.13 (hexane/EtOAc 2 : 1). (2 rotamers 0.8 : 0.2) ^{1}H NMR (300 MHz, CDCl$_3$): δ 7.35 (br s, 0.8 H, CHO), 7.14 (d, *J* = 10 Hz, 0.2 H, CHO), 6.53 (d, *J* = 11.6 Hz, 0.2 H, NH), 6.19 (d, *J* = 1.8 Hz, 0.8 H, NH), 5.51 (dd, *J* = 3.1, 1.2 Hz, 0.8 H, Ar-CH), 5.23 (dd, *J* = 5.2, 3.1 Hz, 0.2 H, Ar-CH), 5.14 (dd, *J* = 4.9, 3.1 Hz, 0.8 H, Ar-CH), 5.03 (dd, *J* = 5.2, 1.2 Hz, 0.8 H, Ar-CH), 4.85 (dd, *J* = 5.2, 1.2 Hz, 0.2 H, Ar-CH), 4.77 (dd, *J* = 3.1, 1.2 Hz, 0.2 H, Ar-CH); ^{13}C NMR (62.5 MHz, CDCl$_3$): δ 163.1 (C), 159.0 (C), 135.2 (C), 134.4 (C), 126.5 (CH), 124.5 (CH), 121.2 (CH), 120.5 (CH), 111.2 (CH), 109.5 (CH); MS (70 eV, EI) m/z (%): 127.1 (100) [M$^+$], 99.1 (39), 72.0 (28); IR (KBr): 3279 (br), 3105, 1653 (C=O), 1539, 1418, 1388, 1208, 773 cm^{-1};

N-(2-Bromothiophen-3-yl)formamide

To a boiling solution of *N*-(thiophen-3-yl)formamide (2.54 g, 20 mmol) in anhydrous chloroform (60 mL) was added NBS (3.52g, 20 mmol) in one portion. After the initial reaction had ceased, the mixture was heated for 10 min, then cooled. The solvent was evaporated under reduced pressure, and the residue was purified by column chromatography on silica gel (hexane/EtOAc 2 : 1, R_f = 0.25) to give 3.27 g (79%) of the title product as a colorless solid, m.p. 92–93 °C. ^{1}H NMR (300 MHz, CDCl$_3$,) (2 rotamers 0.8 : 0.2): δ 8.61 (d, *J* = 11.3 Hz, 0.2 H, CHO), 8.37 (d, *J* = 1.5 Hz, 0.2 H, CHO), 8.19 (br s, 0.2 H, NH), 8.04 (br s, 0.8 H, NH), 7.70 (d, *J* = 6.1 Hz, 0.8 H, Ar-H), 7.32 (d, *J* = 5.8 Hz, 0.2 H, Ar-H), 7.27 (d, *J* = 5.8 Hz, 0.8 H, Ar-H), 6.90 ppm (d, *J* = 6.1 Hz, 0.2 H, Ar-H); ^{13}C NMR (75.5 MHz, CDCl$_3$): δ 158.2 (CH), 157.1 (CH), 134.7 (C), 134.0 (CH), 127.2 (C), 126.7 (C), 125.3 (CH), 122.6 (CH), 119.9 (C),

113.9 (CH); MS (70eV, EI) m/z (%): 207.0 (51) [M$^+$], 205.0 (57) [M$^+$], 179.0 (29), 177.0 (31), 126.1(100), 98.1 (43); IR (KBr): 3220 (br), 1661 (C=O), 1594, 1499, 1393, 1258, 1211, 1001, 823, 708 cm^{-1}; Anal. Calcd for C$_5$H$_4$BrNOS: C 29.14, H 1.96, N 6.80; found: C 28.91, H 1.60, N 6.51

2-Bromo-3-isocyanothiophene (234)

To a solution of *N*-(2-bromothiophen-3-yl)formamide (3.0 g, 14.6 mmol) in anhydrous CH$_2$Cl$_2$ (50 mL) was added at 0 °C triethylamine (8.4 mL, 60.6 mmol), and then dropwise over a period of 10 min POCl$_3$ (1.82 mL, 18.93 mmol). The mixture was stirred at 0 °C for 15 min, then a saturated solution of Na$_2$CO$_3$ (10 mL) was added slowly. The mixture was transferred into a separatory funnel, diluted with CH$_2$Cl$_2$ (100 mL), the organic phase washed with a half-saturated solution of Na$_2$CO$_3$ (50 mL) and brine (50 mL), then dried over anhydrous Na$_2$SO$_4$ and concentrated. The crude product was purified by column chromatography on silica gel (hexane/EtOAc 10 : 1, R_f = 0.31) and subsequent recrystallization from hexane at −18 °C to give 2.33 g (85%) of 2-bromo-3-isocyanothiophene (**234**) as a yellow-red oil. ^{1}H NMR (300 MHz, CDCl$_3$): δ 7.29 (d, *J* = 5.9 Hz, 1 H), 6.98 ppm (d, *J* = 5.9 Hz, 1 H); ^{13}C NMR (125 MHz, CDCl$_3$): δ 168.0 (C), 126.9 (C), 126.3 (CH), 124.6 (CH), 112.1 ppm (C); MS (70eV, EI) m/z (%): 189.0 (100) [M$^+$], 187.0 (98) [M$^+$], 108.1 (45); IR (KBr): 3112, 2120 (NC), 1371, 1010, 950, 713 cm^{-1}; Anal. Calcd for C$_5$H$_2$BrNS: C 31.94, H 1.07, N 7.45; found: C 32.06, H 1.02, N 7.36.

General Procedure for the Reaction of *ortho*-Lithiophenyl (-Hetaryl) Isocyanides with Aldehydes and Ketones (GP8)

To a solution of *o*-bromophenyl (-hetaryl) isocyanide (2 mmol) in anhydrous tetrahydrofuran (20 mL), kept in an oven-dried 25 mL-Schlenk flask under an atmosphere of dry nitrogen, was added dropwise with stirring a 2.5 M solution of *n*-BuLi in hexane (0.8 mL, 2 mmol) at −78 °C over a period of 10 min. The mixture was stirred at −78 °C for another 10 min, before the respective aldehyde (ketone) (2 mmol) in anhydrous THF (2 mL) was added dropwise. The mixture was stirred at −78 °C for 3 h and was then treated in three different ways (variants A–C).

(A) The reaction was quenched with water (2 mL) at −78 °C.

(B) The mixture was gradually warmed to 0 °C within 2 h, and then the reaction was quenched with water (2 mL) at 0 °C.

(C) The mixture was treated with the solution of an electrophile in THF (2 mL) at −78 °C, the resulting mixture stirred at the same temperature for 2 h and warmed to r.t. overnight.

Then the mixture was diluted with diethyl ether (50 mL), washed with water (2 × 10 mL), brine (20 mL) and dried over anhydrous Na_2SO_4. The solvents were removed under reduced pressure to give a crude product, which was purified by column chromatography on silica gel or by Kugelrohr distillation.

(2-Isocyanophenyl)(phenyl)methanol (204a)

The isocyanide **204a** (350 mg, 84%) was obtained from *o*-bromophenyl isocyanide (**159**-Br) (364 mg, 2 mmol) and benzaldehyde (**202a**) (212 mg, 2 mmol) following GP8 (A) and after column chromatography (hexane/ethyl acetate 4 : 1, R_f = 0.27) as a yellow oil. ^{1}H NMR (300 MHz, $CDCl_3$): δ 7.74 (d, *J* = 7.9 Hz, 1 H, Ar-H), 7.47–7.25 (m, 8 H, Ar-H), 6.18 (s, 1 H, CH), 2.51 ppm (s, 1 H, OH); ^{13}C NMR (75.5 MHz, $CDCl_3$, APT): δ 167.6 (C), 141.6 (C), 139.9 (C), 129.7 (CH), 128.7 (2 CH), 128.3 (CH), 128.2 (CH), 127.0 (2 CH), 126.9 (2 CH), 124.4 (C), 71.9 ppm (CH); IR (film): 3393 (br, OH), 3064, 3031, 2896, 2120 (NC), 1483, 1453, 1188, 1035, 1024, 761, 699 cm^{-1}; MS (EI) *m/z* (%): 209 (46) [M^+], 180 (100), 77 (34) ; HRMS (EI): calcd for $C_{14}H_{11}NO^+$ $[M]^+$: 209.0841; found: 209.0839.

(2-Isocyanophenyl)(4-methoxyphenyl)methanol (204b)

The isocyanide **204b** (397 mg, 83%) was obtained from *o*-bromophenyl isocyanide (**159**-Br) (364 mg, 2 mmol) and 4-methoxybenzaldehyde (**202b**) (272 mg, 2 mmol) following GP8 (A) and after column chromatography (hexane/ethyl acetate 5 : 1, R_f = 0.20) as a colorless oil. ^{1}H NMR (300 MHz, $CDCl_3$): δ 7.76 (d, *J* = 7.9 Hz, 1 H, Ar-H), 7.45 (td, *J* = 8.5, 2.3 Hz, 1 H, Ar-H), 7.34–7.25 (m, 4 H, Ar-H), 6.86 (d, *J* = 8.7 Hz, 2 H, Ar-H), 6.11 (d, *J* = 1.9 Hz, 1 H, CH), 3.77 (s, 3 H, CH_3), 2.47 (d, = 2.6 Hz, 1 H, OH); ^{13}C NMR (75.5 MHz, $CDCl_3$): δ 167.5 (C), 159.4 (C), 140.1 (C), 133.7 (C), 129.6 (CH), 128.4 (2 CH), 128.1 (CH), 126.9 (CH), 126.7 (CH), 124.3 (C), 114.0 (2 CH), 71.6 (CH), 55.2 (CH_3); MS (70 eV, EI) *m/z* (%): 239.2 (100) [M^+], 210.2 (92); IR (KBr): 3404 (br, OH), 2933, 2837, 2120 (NC), 1611, 1585, 1511, 1482, 1451, 1304, 1251, 1174, 1112, 1032, 811, 761 cm^{-1}; elemental analysis calcd (%) for $C_{15}H_{13}NO_2$: C 75.30, H. 5.48, N 5.85; found: C 74.98, H. 5.18, N 5.55.

(4-Chlorophenyl)(2-isocyanophenyl)methanol (204c)

The isocyanide **204c** (433 mg, 89%) was obtained from *o*-bromophenyl isocyanide (**159**-Br) (364 mg, 2 mmol) and 4-chlorobenzaldehyde (**202c**) (281 mg, 2 mmol) following GP8 (A) and after column chromatography (hexane/ethyl acetate 5 : 1, R_f = 0.15) as a yellow oil. ^{1}H NMR (300 MHz, CDCl$_3$): δ 7.68 (d, *J* = 7.8 Hz, 1 H, Ar-H), 7.45 (td, *J* = 7.0, 2.2 Hz, 1 H, Ar-H), 7.38–7.29 (m, 6 H, Ar-H), 6.16 (d, *J* = 2.8 Hz, 1 H, CH), 2.48 (d, *J* = 3.4 Hz, 1 H, OH); ^{13}C NMR (75.5 MHz, CDCl$_3$, APT): δ 167.8 (C), 140.0 (C), 139.5 (C), 134.0 (C), 129.9 (CH), 128.8 (2 CH), 128.6 (CH), 128.3 (2 CH), 127.1 (CH), 126.9 (CH), 124.4 (C), 71.2 ppm (CH); MS (70 eV, EI) *m/z* (%): 243 (56) [M$^+$], 214 (84), 180 (100), 77(74); IR (film): 3420 (br), 2361, 2339, 2120 (NC), 1491, 1091, 1035, 1014, 761 cm^{-1}; HRMS (EI): calcd for C$_{14}$H$_{11}$ClNO$^+$ [M]$^+$: 244.05237; found: 244.05243.

(2-Isocyanophenyl)(pyridin-4-yl)methanol (204d)

The isocyanide **204d** (344 mg, 82%) was obtained from *o*-bromophenyl isocyanide (**159**-Br) (364 mg, 2 mmol) and 4-formylpyridine (**202d**) (214 mg, 2 mmol) following GP8 (A) and after column chromatography (dichloromethane/methanol 10 : 1, R_f = 0.18) as a colorless solid, m. p. 139–140 °C. ^{1}H NMR (300 MHz, CDCl$_3$): δ 8.38 (dd, *J* = 4.5, 1.7 Hz, 2 H), 7.60 (d, *J* = 7.8 Hz, 1 H), 7.43 (td, *J* = 7.8, 2.2 Hz, 1 H), 7.38–7.29 (m, 4 H), 6.18 (s, 1 H), 5.42 (br s, 1 H); ^{13}C NMR (75.5 MHz, CDCl$_3$, APT): δ 167.8 (C), 151.6 (2 C), 149.3 (2 CH), 139.2 (C), 130.1 (2 CH), 128.8 (CH), 127.5 (CH), 127.0 (CH), 121.6 (2 CH), 69.8 (CH); MS (70 eV, EI) *m/z* (%): 210.2 (100) [M$^+$], 181.2 (58), 132.1 (34); IR (film): 3037 (br) (OH), 2850 (br), 2123 (NC), 1604, 1416, 1062, 1006, 798, 761 cm^{-1}; elemental analysis calcd (%) for C$_{13}$H$_{10}$N$_2$O: C 74.27, H. 4.79, N 13.33; found: C 73.97, H 4.64, N 13.19.

(2-Isocyanophenyl)(5-methylthiophen-2-yl)methanol (204e)

Compound **204e** (357 mg, 78%) was obtained from *o*-bromophenyl isocyanide (**159**-Br) (364 mg, 2 mmol) and 2-formyl-5-methyl-thiophene (**202e**) (252 mg, 2 mmol) following GP8 (A) and after column chromatography (hexane/ethyl acetate 5 : 1, R_f = 0.24) as a yellow oil. ^{1}H NMR (300 MHz, CDCl$_3$): δ 7.82 (d, *J* = 7.5 Hz, 1 H, Ar-H), 7.50–7.45 (m, 1 H, Ar-H), 7.36–7.32 (m, 2 H, Ar-H), 6.80 (d, *J* = 3.4 Hz, 1 H, thienyl-H), 6.60–6.57 (m, 1 H, thienyl-H), 6.31 (d, *J* = 3.1 Hz, 1 H, OCH),

2.53 (d, J = 3.4 Hz, 1 H, OH), 2.42 ppm (s, 3 H, CH$_3$); ^{13}C NMR (75.5 MHz, CDCl$_3$, APT): δ 167.9 (C), 142.8 (2 C), 141.0 (C), 140.3 (C), 139.4 (C), 129.8 (CH), 128.5 (CH), 126.9 (CH), 126.4 (CH), 126.0 (CH), 124.8 (CH), 68.1 (CH), 15.4 ppm (CH$_3$); MS (DCI) m/z (%): 247.3 (12) [M+NH$_4^+$], 230.2 (100) [M+H$^+$]; IR (KBr): 3403 (br), 2121, 1686, 1482, 1449, 1022, 757 cm^{-1}; HRMS (EI): calcd for C$_{13}$H$_{12}$NOS$^+$ [M+H]$^+$: 230.06396; found: 230.06382.

(2-Isocyanophenyl)(5-methylfuran-2-yl)methanol (204f)

Compound **204f** (377 mg, 88%) was obtained from *o*-bromophenyl isocyanide (**159**-Br) (364 mg, 2 mmol) and 5-methylfuran-2-carb-aldehyde (**202f**) (220 mg, 2 mmol) following GP8 (A) and after column chromatography (hexane/ethyl acetate 5 : 1, R_f = 0.12) as a yellow oil. ^{1}H NMR (300 MHz, CDCl$_3$): δ 7.78 (d, J = 7.9 Hz, 1 H), 7.50–7.44 (m, 1 H), 7.37–7.31 (m, 2 H), 6.11 (d, J = 3.0 Hz, 1 H), 6.01 (d, J = 3.0 Hz, 1 H), 5.90 (dd, J = 3.0, 0.8 Hz, 1 H), 2.63 (d, J = 3.8 Hz, 1 H, OH), 2.26 ppm (s, 3 H, CH$_3$); ^{13}C NMR (75.5 MHz, CDCl$_3$): δ 167.5 (C), 152.9 (C), 151.6 (C), 137.1 (C), 129.6 (CH), 128.6 (CH), 127.4 (CH), 126.8 (CH), 109.15 (C), 109.20 (CH), 106.3 (CH), 65.9 (CH), 13.6 ppm (CH$_3$); MS (EI) m/z (%): 213.1 (52) [M$^+$], 184.1 (38), 170.1 (100); IR (KBr): 3411 (br) (OH), 2121 (NC), 1557, 1449, 1267, 1217, 1200, 1018, 761, 736 cm^{-1}; HRMS (ESI) calcd for C$_{13}$H$_{12}$NO$_2^+$ [M+H$^+$]: 214.08626; found: 214.08644.

1-(2-Isocyanophenyl)-2,2-dimethylpropan-1-ol (204g)

Compound **204g** (303 mg, 80%) was obtained from *o*-bromophenyl isocyanide (**159**-Br) (364 mg, 2 mmol) and 1,1,1-trimethylacetaldehyde (**202g**) (172 mg, 2 mmol) following GP8 (A), after column chromatography (hexane/ethyl acetate 5 : 1, R_f = 0.25) as a yellow oil. ^{1}H NMR (300 MHz, CDCl$_3$): δ 7.60 (d, J = 7.9 Hz, 1 H, Ar-H), 7.41 (t, J = 7.5 Hz, 1 H, Ar-H), 7.36–7.29 (m, 2 H, Ar-H), 4.95 (d, J = 2.3 Hz, 1 H, CH), 2.03 (d, J = 2.6 Hz, 1 H, OH), 0.99 ppm (s, 9 H, *t*Bu); ^{13}C NMR (75.5 MHz, CDCl$_3$, APT): δ 166.9 (C), 138.7 (2 C), 128.9 (CH), 128.7 (CH), 128.0 (CH), 126.7 (CH), 76.2 (CH), 37.2 (C), 25.6 ppm (CH$_3$); MS (DCI) m/z (%): 207 (100) [M+NH$_4^+$], 190 (99) [M+H$^+$]; IR (KBr): 3457 (br) (OH), 2963, 2121 (NC), 1478, 1051, 1006, 757 cm^{-1}; HRMS (ESI) calcd for C$_{12}$H$_{16}$NO$^+$ [M+H$^+$]: 190.12264; found: 190.12267.

1-(2-Isocyanophenyl)-2-methylpropan-1-ol (204h)

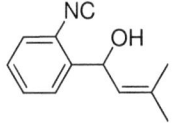

The isocyanide **204h** (127 mg, 36%) was obtained from *o*-bromophenyl isocyanide (**159**-Br) (364 mg, 2 mmol) and isobutyraldehyde (**202h**) (144 mg, 2 mmol) following GP8 (A), after column chromatography (hexane/ethyl acetate 5 : 1, R_f = 0.15) as a yellow oil. ^{1}H NMR (300 MHz, CDCl$_3$): δ 7.57 (d, J = 7.9 Hz, 1 H, Ar-H), 7.43 (td, J = 7.5, 1.5 Hz, 1 H, Ar-H), 7.37–7.27 (m, 2 H, Ar-H), 4.88 (dd, J = 6.2, 3.6 Hz, 1 H, Ar-H), 2.09–1.99 (m, 1 H, CH), 1.99 (d, J = 3.8 Hz, 1 H, CHO), 0.99 (d, J = 6.8 Hz, 3 H, CH$_3$), 0.94 (d, J = 6.8 Hz, 3 H, CH$_3$); ^{13}C NMR (75.5 MHz, CDCl$_3$, APT): δ 166.9 (C), 140.2 (C), 129.5 (CH), 128.0 (CH), 127.4 (CH), 126.7 (CH), 109.2 (C), 74.6 (CH), 34.6 (CH), 19.0 (CH$_3$), 17.2 ppm (CH$_3$); IR (film): 3432 (br, OH), 2963, 2119 (NC), 1450, 1030, 761; MS (EI) *m/z* (%): 175.2 (5) [M$^+$], 132.2 (100); HRMS (EI): *m/z* calcd for C$_{11}$H$_{14}$NO$^+$ [M+H$^+$]: 176.10699; found: 176.10709.

1-(2-Isocyanophenyl)-3-methylbut-2-en-1-ol (204i)

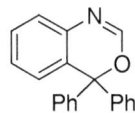

Compound **204i** (260 mg, 70%) was obtained from *o*-bromophenyl isocyanide (**159**-Br) (364 mg, 2 mmol) and 3-methylbut-2-enal (**202i**) (168 mg, 2 mmol) following GP8 (A) (the aldehyde was added at −90 °C), after column chromatography (hexane/ethyl acetate 5 : 1, R_f = 0.18) as a yellow oil. ^{1}H NMR (300 MHz, CDCl$_3$): δ 7.68 (d, J = 7.9 Hz, 1 H, Ar-H), 7.43 (dt, J = 7.2, 1.5 Hz, 1 H, Ar-H), 7.35–7.25 (m, 2 H, Ar-H), 5.81 (dd, J = 9.0, 2.6 Hz, 1 H, OCH), 5.28 (d, J = 9.8, 1 H, CH), 1.96 (d, J = 2.6 Hz, 1 H, OH), 1.92 (s, 3 H, CH$_3$), 1.76 ppm (s, 3 H, CH$_3$); ^{13}C NMR (75.5 MHz, CDCl$_3$, APT): δ 167.0 (C), 140.8 (C), 138.0 (C), 129.7 (CH), 127.9 (CH), 127.0 (CH), 126.6 (CH), 125.4 (CH), 124.0 (C), 66.9 (CH), 25.8 (CH$_3$), 18.8 ppm (CH$_3$); MS (EI) *m/z* (%): 187.2 (4) [M$^+$], 186.2 (25), 77.0 (28), 51.0 (100); IR (KBr): 3389 (br) (OH), 2974, 2914, 2119 (NC), 1483, 1450, 1034, 1006, 762 cm^{-1}; HRMS (ESI) calcd for C$_{12}$H$_{13}$NONa$^+$ [M+Na$^+$]: 210.08894; found: 210.08907.

4,4-Diphenyl-4*H*-3,1-benzoxazine (201k)

Compound **201k** (274 mg, 48%) was obtained from *o*-bromophenyl isocyanide (**159**-Br) (364 mg, 2 mmol) and benzophenone (**202k**) (364 mg, 2 mmol) following GP8 (A) and after column chromatography (hexane/ethyl acetate 4 : 1, R_f = 0.27) as a colorless solid, m. p. 139 °C. ^{1}H NMR (300 MHz, CDCl$_3$): δ 7.36–7.24 (m, 9 H), 7.23–7.18 (m, 4 H), 7.14 (td, J = 7.5, 1.6 Hz, 1 H), 6.69 (d, J = 7.8 Hz, 1 H); ^{13}C NMR (75.5 MHz, CDCl$_3$, APT): δ

150.7 (CH), 142.6 (2 C), 137.7 (C), 129.2 (2 CH), 128.8 (C), 128.3 (2 CH), 128.2 (2 CH), 127.9 (4 CH), 127.1 (2 CH), 126.6 (2 CH), 124.7 (CH), 85.2 (C); MS (EI) m/z (%): 285 (44) [M$^+$], 256 (58), 84 (100); IR (KBr): 1670, 1611, 1595, 1474, 1322, 1289, 1262, 769, 689 cm^{-1}; elemental analysis calcd (%) for C$_{20}$H$_{15}$NO: C 84.19, H 5.30, N 4.91; found: C 83.95, H 5.07, N 5.12.

4-(Trifluoromethyl)-4-phenyl-4*H*-3,1-benzoxazine (201l)

Compound **201l** (403 mg, 78%) was obtained from *o*-bromophenyl isocyanide (**159**-Br) (364 mg, 2 mmol) and 1,1,1-trifluoroacetophenone (**202l**) (348 mg, 2 mmol) following GP8 (A) after column chromatography (hexane/ethyl acetate 5 : 1, R_f = 0.46) and Kugelrohr distillation (0.4 Torr, 100–110 °C) as a colorless oil. ^{1}H NMR (300 MHz, CDCl$_3$): δ 7.51–7.48 (m, 2 H, Ar-H), 7.36–7.34 (m, 4 H, Ar-H), 7.30–7.20 ppm (m, 4 H, Ar-H); ^{13}C NMR (75.5 MHz, CDCl$_3$): δ 147.5 (CH), 136.8 (C), 136.3 (C), 130.5 (CH), 129.4 (CH), 128.4 (2 CH), 127.5 (CH), 127.3 (q, J_{CF} = 1.6 Hz, 2 CH), 126.6 (q, J_{CF} = 2.2 Hz, CH), 126.0 (2 CH), 123.8 (q, J = 287 Hz, C), 121.9 (C), 81.2 ppm (q, J = 31 Hz, C); MS (EI) m/z (%): 277.1 (14) [M$^+$], 208.1 (100); IR (KBr): 3067 (br), 2362, 1632, 1603, 1478, 1458, 1279, 1221, 1101, 985, 943, 765 cm^{-1}; elemental analysis calcd (%) for C$_{15}$H$_{10}$F$_3$NO: C 64.78, H 3.64, N 5.05; found: C 64.98, H 3.57, N 5.39.

4,4-Dimethyl-4*H*-3,1-benzoxazine (201m)

Compound **201m** (166 mg, 52%) was obtained from *o*-bromophenyl isocyanide (**159**-Br) (364 mg, 2 mmol) and acetone (**202m**) (116 mg, 2 mmol) following GP8 (A) and after column chromatography (hexane/ethyl acetate 5 : 1, R_f = 0.20) as a colorless oil. ^{1}H NMR (300 MHz, CDCl$_3$): δ 7.28–7.13 (m, 4 H, Ar-H, CH=N), 7.07 (d, J = 7.5 Hz, 1 H, Ar-H), 1.64 (s, 6 H, CH$_3$) ppm; ^{13}C NMR (75.5 MHz, CDCl$_3$, APT): δ 150.8 (CH), 136.5 (C), 131.7 (C), 128.4 (CH), 127.1 (CH), 124.8 (CH), 122.5 (CH), 77.8 (C), 29.1 (CH$_3$); MS (EI) m/z (%): 161.2 (24), 146.2 (100); IR (film): 2980, 1621, 1485, 1452, 1366, 1227, 1133, 1105, 1080, 768 cm^{-1}; HRMS (EI): m/z calcd for C$_{10}$H$_{11}$NO$^+$ [M$^+$]: 161.0841; found: 161.0840.

(2-Isocyanophenyl)(pyridin-4-yl)methyl methyl carbonate (205)

Compound **205** (301 mg, 56%) was obtained from *o*-bromophenyl isocyanide (**159**-Br) (376 mg, 2 mmol), 4-formylpyridine (**202d**) (214 mg, 2 mmol) and methyl chloroformate (189 mg, 2 mmol) following GP8 (C), and after column chromatography on silica gel (hexane/ethyl acetate 1 : 1, R_f = 0.33) as a yellow oil. ^{1}H NMR (300 MHz, CDCl$_3$): δ 8.62 (dd, *J* = 4.4, 1.6 Hz, 2 H, Ar-H), 7.52–7.39 (m, 4 H, Ar-H), 7.34 (dd, *J* = 4.4, 1.6 Hz, 2 H, Ar-H), 6.97 (s, 1 H, CH), 3.83 ppm (s, 3 H, CH$_3$); ^{13}C NMR (75.5 MHz, CDCl$_3$): δ 169.3 (C), 154.4 (C), 150.3 (2 CH), 146.1 (C), 134.7 (C), 130.0 (CH), 129.7 (CH), 127.5 (2 CH), 124.7 (C), 121.4 (2 CH), 74.9 (CH), 55.5 ppm (CH$_3$); MS (EI) *m/z* (%): 268.2 (48) [M$^+$], 209.1 (100); IR (KBr): 3032, 2958, 2120 (NC), 1751, 1599, 1441, 1259, 984, 950, 764 cm^{-1}; HRMS (ESI) calcd for C$_{15}$H$_{13}$N$_2$O$_3$$^+$ [M+H$^+$]: 269.09262; found: 269.09268.

3-(Trifluoromethyl)-3-methylisobenzofuran-1(3*H*)-imine (210o)

The compound **210o** (250 mg, 58%) was obtained from *o*-bromophenyl isocyanide (**159**-Br) (364 mg, 2 mmol) and 1,1,1-trifluoroacetone (**202o**) (224 mg, 2 mmol) following GP8 (B) and after Kugelrohr distillation (0.1 Torr, 85–95 °C) as a colorless oil. ^{1}H NMR (300 MHz, CDCl$_3$): δ 7.39 (td, *J* = 7.2, 1.9 Hz, 1 H, Ar-H), 7.30–7.21 (m, 3 H, Ar-H), 7.15 (s, 1 H, NH), 1.87 (s, 3 H, CH$_3$); ^{13}C NMR (125 MHz, CDCl$_3$, APT): δ 148.4 (C), 136.7 (C), 130.7 (CH), 127.9 (CH), 125.8 (CH), 125.3 (CH), 124.1 (q, J_{CF} = 287 Hz, C), 121.4 (C), 77.2 (q, J^2_{CF} = 31 Hz, C), 22.0 ppm (CH$_3$); MS (EI) *m/z* (%): 215.0 (39) [M$^+$], 146.0 (100); IR (KBr): 3304 (br) (NH), 1676, 1636, 1456, 1295, 1225, 1179, 1096, 769 cm^{-1}; elemental analysis calcd (%) for C$_{10}$H$_8$F$_3$NO: C 55.82, H 3.75, N 6.51; found: C 55.74, H 3.51, N 6.30.

Methyl 4-(Trifluoromethyl)-4-phenyl-4*H*-3,1-benzoxazine-2-carboxylate (201l-CO$_2$Me)

Compound **201l**-CO$_2$Me (302 mg, 45%) was obtained from *o*-bromophenyl isocyanide (**159**-Br) (376 mg, 2 mmol), 1,1,1-trifluoroacetophenone (**202l**) (348 mg, 2 mmol) and methyl chloroformate (189 mg, 2 mmol) following GP8 (C), and after column chromatography on silica gel (hexane/ethyl acetate 4 : 1, R_f = 0.26) as a yellow oil. ^{1}H NMR (300 MHz, CDCl$_3$): δ 7.50 (m, 4 H, Ar-H), 7.39 (m, 5 H, Ar-H), 3.98 ppm (s, 3 H, CH$_3$); ^{13}C NMR (125 MHz, CDCl$_3$, APT): δ 159.3

(C), 145.6 (C), 137.0 (C), 135.0 (C), 130.7 (CH), 129.7 (CH), 129.3 (CH), 128.4 (CH), 127.4 (CH), 127.3 (CH), 126.1 (q, J = 2.3 Hz, CH), 123.4 (q, J = 285.4 Hz, C), 121.1 (C), 83.0 (q, J = 30.8 Hz, C), 53.8 ppm (CH$_3$); MS (EI) m/z (%): 335 (16) [M$^+$], 266 (88), 43 (100); IR (KBr): 2955, 1745, 1647, 1601, 1327, 1293, 1211, 1180, 990, 767, 702 cm^{-1}; HRMS (ESI) calcd for C$_{17}$H$_{12}$NF$_3$O$_3$Na$^+$ [M+Na$^+$]: 358.0661; found: 358.0666.

Ethyl 2-(4-(Trifluoromethyl)-4-phenyl-4H-3,1-benzoxazin-2-yl)acetate (20 1l-CH$_2$CO$_2$Et)

Compound **20 1l-CH$_2$CO$_2$Et** (338 mg, 47%) was obtained from o-bromophenyl isocyanide (**159**-Br) (376 mg, 2 mmol) and 1,1,1-trifluoroacetophenone (**202l**) (348 mg, 2 mmol) following GP8 (C) and after column chromatography on silica gel (hexane/ethyl acetate 5 : 1, R_f = 0.20) as a colorless solid, m.p. 56–57 °C. ^{1}H NMR (300 MHz, CDCl$_3$): δ 7.55–7.52 (m, 2 H, Ar-H), 7.43–7.36 (m, 5 H, Ar-H), 7.30–7.23 (m, 2 H, Ar-H), 4.19–3.95 (m, 4 H, CH$_2$), 1.73 ppm (s, 3 H, CH$_3$); ^{13}C NMR (125 MHz, CDCl$_3$, APT): δ 157.0 (C), 137.7 (C), 136.2 (C), 130.5 (2 CH), 129.5 (CH), 128.4 (2 CH), 127.54 (CH), 127.45 (CH), 126.5 (CH), 126.4 (CH), 123.9 (q, J_{CF} = 287 Hz, C), 120.4 (C), 105.8 (C), 82.0 (q, J^2_{CF} = 31 Hz, C), 65.8 (CH$_2$), 65.4 (CH$_2$), 22.3 ppm (CH$_3$); MS (EI) m/z (%): 363.2 (16) [M$^+$], 320.2 (44), 87.1 (100); IR (KBr): 2991, 2911, 1657, 1484, 1454, 1247, 1181, 1165, 1119, 1028, 949, 777 cm^{-1}; HRMS (ESI) calcd for C$_{19}$H$_{17}$F$_3$NO$_3^+$ [M+H$^+$]: 364.11550; found: 364.11561.

4-(Trifluoromethyl)-4-phenyl-1H-3,1-benzoxazin-2(4H)-one (206)

Compound **206** (450 mg, 77%) was obtained from o-bromophenyl isocyanide (**159**-Br) (376 mg, 2 mmol), 1,1,1-trifluoroacetophenone (**202l**) (348 mg, 2 mmol) and iodine (508 mg, 2 mmol) following GP8 (C) and after column chromatography on silica gel (hexane/ethyl acetate 5 : 1, R_f = 0.12) as a colorless solid, m.p. 158–159 °C. ^{1}H NMR (300 MHz, CDCl$_3$): δ 9.49 (s, 1 H, NH), 7.50–7.34 (m, 7 H, Ar-H), 7.19 (dt, J = 7.9, 1.1 Hz, 1H, Ar-H), 6.98 ppm (d, J = 8.3 Hz, 1 H, Ar-H); ^{13}C NMR (75.5 MHz, CDCl$_3$, APT): δ 150.7 (C), 135.0 (C), 134.3 (C), 130.7 (CH), 129.9 (CH), 128.6 (2 CH), 127.4 (2 CH), 126.1 (q, J_{CF} = 2.8 Hz, CH), 123.7 (CH), 123.2 (q, J_{CF} = 285 Hz, C), 116.8 (C), 115.7 (CH), 85.5 ppm (q, J^2_{CF} = 31.4 Hz, C); MS (DCI) m/z (%): 604.4 (60) [2M+Na$^+$], 328.2 (88) [M+NH$_3$+NH$_4^+$], 311.2 (100) [M+NH$_4^+$]; IR (KBr): 3100 (NH), 1724 (C=O), 1599, 1498, 1365, 1283, 1174, 1057, 984, 790, 764, 722 cm^{-1}; elemental analysis calcd (%) for C$_{15}$H$_{10}$F$_3$NO$_2$: C 61.44, H 3.44, N 4.78; found: C 61.16, H 3.17, N 5.02.

4-(Trifluoromethyl)-2-morpholino-4-phenyl-4*H*-3,1-benzoxazine (207)

Compound **207** (395 mg, 55%) was obtained from *o*-bromophenyl isocyanide (**159**-Br) (376 mg, 2 mmol), 1,1,1-trifluoroacetophenone (**202l**) (348 mg, 2 mmol) and iodine (508 mg, 2 mmol) following GP8 (C) [morpholine (348 mg, 4 mmol) was added, and the mixture was stirred at r.t. for 1 h before aqueous work up] and after column chromatography on silica gel (hexane/ethyl acetate 2 : 1, R_f = 0.30) as a colorless solid, m.p. 106–107 °C. ^{1}H NMR (300 MHz, CDCl$_3$): δ 7.41–7.34 (m, 5 H, Ar-H), 7.32 (dt, *J* = 7.8, 1.6 Hz, 1H, Ar-H), 7.26–7.23 (m, 1 H, Ar-H), 7.05 (d, *J* = 7.8 Hz, 1 H, Ar-H), 7.04 (dt, *J* = 6.9, 1.3 Hz, 1 H, Ar-H), 3.75–3.65 ppm (m, 8 H, CH$_2$); ^{13}C NMR (75.5 MHz, CDCl$_3$, APT): δ 151.0 (C), 142.3 (C), 135.6 (C), 130.4 (CH), 129.5 (CH), 128.4 (2 CH), 127.5 (2 CH), 125.9 (C), 125.5 (q, J_{CF} = 2.3 Hz, CH), 123.3 (CH), 122.6 (CH), 122.1 (C), 118.7 (C), 66.5 (CH$_2$), 44.6 ppm(CH$_2$); MS (DCI) *m/z* (%): 363.3 (100) [M+H$^+$]; IR (KBr): 2862, 1634, 1592, 1483, 1426, 1290, 1254, 1168, 1118, 1072, 1029, 988, 924, 862, 766, 719, 653 cm^{-1}; elemental analysis calcd (%) for C$_{19}$H$_{17}$F$_3$N$_2$O$_2$: C 62.98, H 4.73, N 7.73; found: C 62.66, H 4.53, N 7.90

General Procedure for the Cu$_2$O-Catalyzed Cyclization of (2-Isocyanophenyl)methanols 204 (GP9)

To a solution of isocyanobenzylalcohol **204** (2 mmol) in benzene (10 mL) was added Cu$_2$O (14.4 mg, 5 mol%), and the resulting mixture was heated under reflux for 1 h. Then, the mixture was cooled to r.t., the solvent was removed under reduced pressure, and the product was purified by column chromatography on silica gel.

4-Phenyl-4*H*-3,1-benzoxazine (201a)

Compound **201a** (301 mg, 86%) was obtained from isocyanobenzylalcohol **204a** (350 mg, 1.67 mmol) following GP9, and after column chromatography on silica gel (hexane/ethyl acetate/Et$_3$N 5 : 1 : 1, R_f = 0.38) as a slightly yellow solid, m.p. 62–63 °C. ^{1}H NMR (300 MHz, CDCl$_3$): δ 7.40–7.22 (m, 7 H, Ar-H), 7.19 (s, 1 H, CH=N), 7.12 (dt, *J* = 7.5, 1.5 Hz, 1 H, Ar-H), 6.73 (d, *J* = 7.5 Hz, 1 H, Ar-H), 6.29 ppm (s, 1 H, CH); ^{13}C NMR (75.5 MHz, CDCl$_3$): δ 150.4 (CH), 139.8 (C), 137.1 (C), 129.1 (CH), 129.0 (CH), 128.7 (CH), 127.8 (CH), 127.1 (CH), 125.5 (CH), 125.3 (C), 124.8 (CH), 77.4 ppm (CH); MS (EI) *m/z* (%): 209.0 (56) [M$^+$], 180.0 (100); IR (KBr): 1612, 1601, 1489, 1455, 1219, 1125, 1096, 773 cm^{-1}; elemental analysis calcd (%) for C$_{14}$H$_{11}$NO: C 80.36, H 5.30, N 6.69; found: C 80.71, H 5.19, N 6.88.

4-(4-Methoxyphenyl)-4*H*-3,1-benzoxazine (201b)

Compound **201b** (222 mg, 74%) was obtained from isocyanobenzylalcohol **204b** (300 mg, 1.26 mmol) following GP9, and after column chromatography on silica gel (hexane/ethyl acetate/Et$_3$N 5 : 1 : 1, R_f = 0.29) as a yellow oil. ^{1}H NMR (300 MHz, CDCl$_3$): δ 7.32–7.22 (m, 4 H, Ar-H), 7.17 (s, 1 H, CH=N), 7.13 (dt, J = 7.5, 1.5 Hz, 1 H, Ar-H), 6.90 (d, J = 8.7 Hz, 2 H, Ar-H), 6.73 (d, J = 7.5 Hz, 1 H, Ar-H), 6.26 (s, 1 H, CH), 3.80 ppm (s, 3 H, CH$_3$); ^{13}C NMR (75.5 MHz, CDCl$_3$): δ 160.1 (C), 150.5 (CH), 137.3 (C), 132.1 (C), 129.4 (2 CH), 129.0 (CH), 127.1 (CH), 125.5 (CH), 125.5 (C), 124.7 (CH), 114.1 (2 CH), 77.1 (CH), 55.3 ppm (CH$_3$); MS (EI) *m/z* (%): 239.0 (100) [M$^+$], 210.0 (99); IR (KBr): 2957, 2933, 1605, 1510, 1455, 1250, 1175, 1122, 1092, 1032, 827, 770 cm^{-1}; HRMS (ESI) calcd for C$_{15}$H$_{14}$NO$_2^+$ [M+H$^+$]: 240.10191; found: 240.10206.

4-(4-Chlorophenyl)-4*H*-3,1-benzoxazine (201c)

Compound **201c** (331 mg, 75%) was obtained from isocyanobenzylalcohol **204c** (443 mg, 1.82 mmol) following GP9, and after column chromatography on silica gel (hexane/ethyl acetate/Et$_3$N 5 : 1 : 1, R_f = 0.40) as a yellow oil. ^{1}H NMR (300 MHz, CDCl$_3$): 7.37–7.22 (m, 6 H, Ar-H), 7.18 (s, 1 H, CH=N), 7.14 (dt, J = 7.5, 1.5 Hz, 1 H, Ar-H), 6.72 (d, J = 7.5 Hz, 2 H, Ar-H), 6.27 ppm (s, 1 H, CH); ^{13}C NMR (75.5 MHz, CDCl$_3$): δ 150.1 (CH), 138.2 (C), 137.0 (C), 135.0 (C), 129.3 (CH), 129.2 (2 CH), 129.0 (2 CH), 127.3 (CH), 125.4 (CH), 125.0 (CH), 124.8 (C), 76.6 ppm (CH); MS (EI) *m/z* (%): 245 (17) [M$^+$+2], 243 (60) [M$^+$], 216 (30), 214 (100), 180 (68%); IR (KBr): 3051, 1610, 1485, 1457, 1215, 1120, 1085, 1016, 832, 770 cm^{-1}; HRMS (ESI) calcd for C$_{14}$H$_{11}$NOCl$^+$ [M+H$^+$]: 244.05237; found: 244.05252.

4-(Pyridin-4-yl)-4*H*-3,1-benzoxazine (201d)

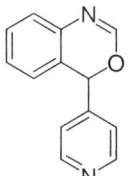

Compound **201d** (252 mg, 73%) was obtained from isocyanobenzylalcohol **204d** (346 mg, 1.20 mmol) following GP9, and after column chromatography on silica gel (ethyl acetate/Et$_3$N 20 : 1, R_f = 0.40) as a colorless solid, m.p. 76–77 °C. ^{1}H NMR (300 MHz, CDCl$_3$): δ 8.64 (dd, *J* = 4.1, 1.5 Hz, 2 H, Ar-H), 7.34 (dt, *J* = 7.9, 1.5 Hz, 1 H, Ar-H), 7.27–7.24 (m, 4 H, Ar-H, CH=N), 7.19 (dt, *J* = 7.5, 1.5 Hz, 1 H, Ar-H), 6.79 (d, *J* = 7.5 Hz, 1 H, Ar-H), 6.27 ppm (s, 1 H, CH); ^{13}C NMR (75.5 MHz, CDCl$_3$): δ 150.3 (2 CH), 149.8 (CH), 147.8 (C), 136.7 (C), 129.6 (CH), 127.4 (CH), 125.1 (CH), 123.6 (C), 121.9 (2 CH), 75.6 ppm (CH); MS (EI) *m/z* (%): 210 (90) [M$^+$], 181 (100), 132 (28); IR (KBr): 1612, 1557, 1485, 1452, 1411, 1393, 1326, 1268, 1220, 1130, 1098, 960, 918, 834, 784, 650, 605 cm^{-1}; elemental analysis calcd (%) for C$_{13}$H$_{10}$N$_2$O: C 74.27, H 4.79; found: C 74.26, H 4.98.

4-*tert*-Butyl-4*H*-3,1-benzoxazine (201g)

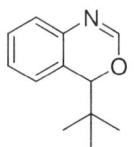

Compound **201g** (152 mg, 83%) was obtained from isocyanobenzylalcohol **204g** (184 mg, 0.97 mmol) following GP9, and after column chromatography on silica gel (hexane/Et$_3$N 10 : 1, R_f = 0.36) as a colorless oil. Alternatively, **201g** was obtained with KO*t*Bu as a catalyst: To the solution of **204g** (100 mg, 0.53 mmol) in dichloromethane (5 mL) was added at r.t. KO*t*Bu (12 mg, 0.11 mmol). The mixture was stirred for 2 h, diluted with dichloromethane (20 mL), washed with water (2 × 5 mL), the organic phase was dried over Na$_2$SO$_4$, filtrated and concentrated under reduced pressure to give a crude product, which was purified by column chromatography on silica gel (hexane/Et$_3$N 10 : 1, R_f = 0.36) to give 65 mg (65%) of **201g** as a colorless oil. ^{1}H NMR (300 MHz, CDCl$_3$): δ 7.28 (dt, *J* = 7.5, 2.3 Hz, 1 H, Ar-H), 7.21 (s, 1 H, CH=N), 7.18–7.13 (m, 2 H, Ar-H), 7.90 (d, *J* = 7.5 Hz, 1 H, Ar-H), 4.92 (s, 1 H, CH), 0.97 ppm (s, 9 H, *t*Bu); ^{13}C NMR (75.5 MHz, CDCl$_3$, APT): δ 151.5 (CH), 138.1 (C), 128.8 (CH), 126.7 (CH), 126.0 (CH), 124.5 (CH), 122.8 (C), 83.7 (CH), 38.8 (C), 25.1 ppm (CH$_3$); MS (DCI) *m/z* (%): 207.2 (4) [M+NH$_4^+$], 189.2 (14) [M$^+$], 132.1 (100), 122.1 (26); IR (KBr): 3281, 2957, 1695, 1621, 1479, 1218, 1124, 1098, 767 cm^{-1}; HRMS (ESI) calcd for C$_{12}$H$_{16}$NO$^+$ [M+H$^+$]: 190.12264; found: 190.12263.

3-(5-Methylfuran-2-yl)isobenzofuran-1(3*H*)-imine (210f)

Compound **210f** (165 mg, 66%) was obtained from isocyanobenzylalcohol **204f** (250 mg, 1.17 mmol) following GP9 and after column chromatography (hexane/ethyl acetate/triethylamine 5 : 1 : 1, R_f = 0.59) as a yellow oil. ^{1}H NMR (300 MHz, CDCl$_3$): δ 7.32 (dt, *J* = 7.9, 1.5 Hz, 1 H, Ar-H), 7.23 (d, *J* = 7.9, 1.1 Hz, 1 H, Ar-H), 7.18 (dt, *J* = 7.5, 1.5 Hz, 1 H, Ar-H), 7.14 (s, 1 H, NH), 6.92 (d, *J* = 7.5 Hz, 1 H, Ar-H), 6.29 (s, 1 H, CH), 6.03 (d, *J* = 3.4 Hz, 1 H, furyl-H), 5.92 (m, 1 H, furyl-H), 2.29 ppm (s, 3 H, CH$_3$); ^{13}C NMR (125 MHz, CDCl$_3$, APT): δ 153.7 (C), 150.3 (C), 149.9 (C), 137.3 (C), 129.3 (CH), 126.9 (CH), 125.3 (CH), 124.9 (CH), 122.8 (C), 111.4 (CH), 106.4 (CH), 70.1 (CH), 13.7 ppm (CH$_3$); IR (film): 3423 (br, NH), 1621 (C=N), 1215, 1121, 908, 729 cm^{-1}; MS (EI) *m/z* (%): 213.0 (28) [M$^+$], 184.0 (44), 170.0 (100); HRMS (ESI) calcd for C$_{13}$H$_{12}$NO$_2^+$ [M+H$^+$]: 214.08626; found: 214.08638.

3-Isopropylisobenzofuran-1(3*H*)-imine (210h)

Compound **210h** (68 mg, 68%) was obtained from isocyanobenzylalcohol **204h** (100 mg, 0.57 mmol) following GP9 and after column chromatography (hexane/ethyl acetate/triethylamine 5 : 1 : 0.5, R_f = 0.40) as a yellow oil. ^{1}H NMR (300 MHz, CDCl$_3$): δ 7.24 (dd, *J* = 7.5, 1.9 Hz, 1 H, Ar-H), 7.17 (s, 1 H, NH), 7.19–7.12 (m, 2 H, Ar-H), 6.89 (d, *J* = 6.8 Hz, 1 H), 5.12 (d, *J* = 4.1 Hz, 1 H, C(3)-H), 2.15–2.02 (m, 1 H, iPr-CH), 1.03 (d, *J* = 6.8 Hz, 3 H, CH$_3$), 0.95 ppm (d, *J* = 6.8 Hz, 3 H, CH$_3$); ^{13}C NMR (125 MHz, CDCl$_3$, APT): δ 151.0 (C), 137.4 (C), 128.6 (CH), 126.6 (CH), 125.0 (C), 124.7 (CH), 124.5 (CH), 80.5 (CH), 35.3 (CH), 18.5 (CH$_3$), 16.2 ppm (CH$_3$); IR (film): 3422 (br, NH), 2965, 1624, 1130, 765 cm^{-1}; MS (EI) *m/z* (%): 175.1 (17) [M$^+$], 132.0 (100); HRMS (EI): *m/z* calcd for C$_{11}$H$_{14}$NO$^+$ [M+H$^+$]: 176.10699; found: 176.10701.

6-(Pyridin-4-yl)thieno[3,2-c]furan-4(6*H*)-imine (211d)

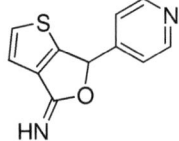

The crude isocyanobenzylalcohol **212d** was obtained from 2-bromo-3-isocyanothiophene (**234**) (376 mg, 2 mmol) and pyridine-4-carbaldehyde (**202d**) (214 mg, 2 mmol) following GP8 (A) as a yellow oil (TLC: hexane/ethyl acetate 5 : 1, R_f = 0.15). Compound **211d** (320 mg, 74% over two steps) was obtained following GP2 and after column chromatography on silica gel (ethyl acetate/triethylamine 15 : 1, R_f = 0.38) as a colorless solid, m.p. 133–134 °C. ^{1}H NMR (300 MHz,

DMSO[d6]): δ 8.57 (dd, J = 4.1, 1.5 Hz, 2 H), 7.59 (d, J = 5.3 Hz, 1 H), 7.39 (dd, J = 4.5, 1.9 Hz, 2 H), 7.19 (d, J = 5.3 Hz, 1 H), 6.82 (d, J = 4.5 Hz, 1 H), 6.08 ppm (d, J = 4.1 Hz, 1 H); ^{13}C NMR (125 MHz, DMSO[d6], APT): δ 166.5 (C), 150.7 (C), 149.6 (2 CH), 145.6 (C), 125.9 (CH), 124.7 (CH), 120.9 (2 CH), 118.7 (C), 67.1 ppm (CH); MS (EI) m/z (%): 216.0 (100) [M$^+$], 187.0 (24); IR (KBr): 2824 (br) (NH), 2118, 1603, 1415, 1270, 1066, 1008, 966, 720, 695, 617 cm^{-1}; elemental analysis calcd (%) for $C_{11}H_8N_2OS$: C 61.09, H 3.73, N 12.95; found: C 60.97, H 3.57, N 12.78.

3-(Pyridin-2-yl)indolin-2-one (215n)

The compound **215n** (330 mg, 79%) was obtained from *o*-bromophenyl isocyanide (**159**-Br) (364 mg, 2 mmol) and pyridyl-2-carbaldehyde (**202n**) (214 mg, 2 mmol) following GP8 (A) and after column chromatography [hexane/ethyl acetate 1 : 1 to ethyl acetate, R_f = 0.10 (1 : 1)] as a colorless solid, m. p. 98−99 °C. ^{1}H NMR (300 MHz, CDCl$_3$, 2 rotamers 0.4 : 0.6): δ 9.51 (m, 1 H), 8.52 (d, J = 5.0 Hz, 1 H), 8.40 (d, J = 11.5 Hz, 0.4 H), 8.26 (d, J = 1.9 Hz, 0.6 H), 8.03 (d, J = 8.1 Hz, 0.6 H), 7.66 (dt, J = 7.8, 1.9 Hz, 1 H), 7.45 (dd, J = 7.8, 1.9 Hz, 0.4 H), 7.37−7.11 (m, 5 H), 5.92 (s, 0.4 H), 5.88 (s, 0.6 H), 4.64 ppm (br s, 1 H); ^{13}C NMR (75.5 MHz, CDCl$_3$, APT): δ 162.4 (CH), 160.7 (C), 160.4 (C), 159.1 (CH), 148.1 (CH), 147.7 (CH), 137.5 (CH), 137.4 (CH), 135.5 (C), 135.2 (C), 133.4 (C), 131.7 (C), 129.0 (CH), 128.8 (CH), 128.3 (CH), 128.1 (CH), 125.4 (CH), 124.8 (CH), 123.7 (CH), 122.8 (CH), 122.7 (CH), 120.7 (CH), 120.5 (CH), 119.9 (CH), 74.0 (CH), 73.3 ppm (CH); MS (EI) m/z (%): 210.1 (74) [M$^+$], 181.1 (100), 132.1 (68); IR (KBr): 3332 (br), 1682, 1590, 1520, 1453, 1437, 1300, 1267, 1059, 757, 732 cm^{-1}; HRMS (EI): m/z calcd for $C_{13}H_{11}N_2O^+$ [M+H$^+$]: 211.08659; found: 211.08654.

3,3-Diphenylindolin-2-one (215k)[197]

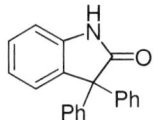

Compound **215k** (239 mg, 42%) was obtained from *o*-bromophenyl isocyanide (**159**-Br) (364 mg, 2 mmol) and benzophenone (**202k**) (364 mg, 2 mmol) following GP8 (B) and after column chromatography on silica gel (hexane/ethyl acetate 4 : 1, R_f = 0.11) as a colorless solid, m. p. 224−225 °C (lit.[197] 227−228 °C). ^{1}H NMR (300 MHz, CDCl$_3$): δ 9.20 (br s, 1 H, NH), 7.30−7.19 (m, 12 H, Ar-H), 7.03 (t, J = 7.2 Hz, 1 H, Ar-H), 6.95 ppm (d, J = 7.2 Hz, 1 H, Ar-H); ^{13}C NMR (75.5 MHz, CDCl$_3$): δ 180.3 (C), 141.6 (2 C), 140.3 (C), 133.5 (C), 128.4 (8 CH), 128.2 (CH), 127.3 (2 CH), 126.2 (CH), 122.8 (CH), 110.5 (CH), 63.1 ppm (C); MS (EI) m/z (%): 285.1 (100) [M$^+$], 256.1 (74); IR (KBr): 3250 (br) (NH), 1725, 1683, 1472, 1322, 1204, 742, 698, 609 cm^{-1}; elemental analysis calcd (%) for C$_{20}$H$_{15}$NO: C 84.19, H 5.30, N 4.91; found: C 84.40, H 5.46, N 5.07.

6,6-Diphenyl-4*H*-thieno[3,2-b]pyrrol-5(6*H*)-one (217k)

Compound **217k** (303 mg, 52%) was obtained from 2-bromo-3-isocyanothiophene (**234**) (376 mg, 2 mmol) and benzophenone (**202k**) (364 mg, 2 mmol) following GP8 (B) and after column chromatography on silica gel (hexane/ethyl acetate 4 : 1, R_f = 0.13) as a colorless solid, m.p. 197–198 °C. ^{1}H NMR (300 MHz, CDCl$_3$): δ 9.38 (br s, 1 H, NH), 7.36–7.22 (m, 11 H, Ar-H), 6.76 ppm (d, J = 5.1 Hz, 1 H, Ar-H); ^{13}C NMR (125 MHz, CDCl$_3$, APT): δ 182.8 (C), 141.6 (2 C), 141.1 (C), 128.6 (4 CH), 128.0 (CH), 128.1 (4 CH), 127.4 (2 CH), 125.6 (C), 112.8 (CH), 64.2 ppm (C); MS (EI) *m/z* (%): 291.2 (58) [M$^+$], 262.2 (100); IR (KBr): 3023 (br) (NH), 1705 (C=O), 1493, 1270, 1092, 836, 756, 696 cm^{-1}; elemental analysis calcd (%) for C$_{18}$H$_{13}$NOS: C 74.20, H 4.50, N 4.81; found: C 74.05, H 4.32, N 4.77.

6-(Trifluoromethyl)-6-phenylthieno[2,3-c]furan-4(6*H*)-imine (217l)

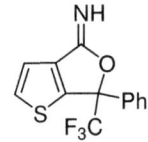

Compound **217l** (425 mg, 75%) was obtained from 2-bromo-3-isocyanothiophene (**234**) (376 mg, 2 mmol) and 1,1,1-trifluoroacetophenone (**202l**) (348 mg, 2 mmol) following GP8 (B) and after column chromatography on silica gel (hexane/ethyl acetate 10 : 1, R_f = 0.13) as a yellow oil. ^{1}H NMR (300 MHz, CDCl$_3$): δ 7.65–7.62 (m, 2 H), 7.46–7.37 (m, 3 H), 7.32 (d, J = 5.3 Hz, 1 H), 7.26 (s, 1 H), 6.98 ppm (d, J = 5.3 Hz, 1 H); ^{13}C NMR (125 MHz, CDCl$_3$, APT): δ 147.1 (C), 140.0 (C), 136.0 (C), 129.7 (2 CH), 128.7 (2 CH), 126.4 (CH), 126.3 (CH), 124.5 (CH), 123.2 (q, J_{CF} = 287 Hz, C), 116.3 (C), 82.1 ppm (q, J^2_{CF} = 33 Hz, C); MS (EI) *m/z* (%): 283.2 (16) [M$^+$], 254.1 (58), 214.1 (100); IR (KBr): 3067, 1613, 1293, 1181, 1086, 955, 741 cm^{-1}; HRMS (EI): *m/z* calcd for C$_{13}$H$_9$F$_3$NOS$^+$ [M+H$^+$]: 284.03515; found: 284.03515.

7-(Trifluoromethyl)-7-phenylfuro[3,4-b]pyridin-5(7H)-imine (219l)

Compound **219l** (179 mg, 64%) was obtained from 2-bromo-3-isocyanopyridine (183 mg, 1 mmol) and 1,1,1-trifluoroacetophenone (**202l**) (174 mg, 1 mmol) following GP8 (B) and after column chromatography on silica gel (hexane/ethyl acetate 5 : 1, R_f = 0.18) as a colorless solid, m. p. 82–83 °C. ^{1}H NMR (300 MHz, CDCl$_3$): δ 8.57 (dd, J = 4.8, 1.8 Hz, 1 H), 7.69–7.66 (m, 2 H), 7.55 (dd, J = 7.7, 1.5 Hz, 1 H), 7.40–7.33 ppm (m, 5 H); ^{13}C NMR (125 MHz, CDCl$_3$, APT): δ 148.31 (C), 148.25 (CH), 141.3 (C), 134.6 (C), 133.4 (C), 132.9 (CH), 129.6 (2 CH), 128.4 (2 CH), 127.1 (CH), 125.4 (CH), 123.2 (q, J_{CF} = 287 Hz, C), 81.3 ppm (q, J^2_{CF} = 30 Hz, C); MS (EI) m/z (%): 278.2 (54) [M$^+$], 109.2 (100), 181.2 (43), 105.1 (58); IR (KBr): 3435 (br) (NH), 1635, 1169 cm^{-1}; elemental analysis calcd (%) for C$_{14}$H$_9$F$_3$N$_2$O: C 60.44, H 3.26, N 10.07; found: C 60.15, H 3.12, N 9.89.

General Procedure for the Synthesis of 4H-3,1-Benzoxazine-4-ones 199-Nu and Isatoic Anhydride (209) (GP10)

To a solution of *o*-bromophenyl isocyanide (**159**-Br) (364 mg, 2 mmol) in anhydrous THF (20 mL), kept in an oven-dried 25 mL-Schlenk flask under an atmosphere of dry nitrogen, was added dropwise with stirring a 2.5 M solution of *n*-BuLi in hexane (0.8 mL, 2 mmol) at –78 °C over a period of 10 min. The mixture was stirred at –78 °C for 10 min, and CO$_2$ was bubbled through the mixture at –78 °C for 1 min. The mixture was stirred at –78 °C for 1h, then a solution of I$_2$ (508 mg, 2 mmol) in anhydrous THF (2 mL) was added dropwise, and the temperature was allowed to rise to 20 °C over a period of 1 h. Water (for the synthesis of **209**) or the solution of the corresponding amine (2 mmol) and triethylamine (2 mmo) in THF (2 mL) was added, and the mixture was stirred at r.t. for 2 h After addition of saturated NH$_4$Cl solution (20 mL), the mixture was diluted with diethyl ether (50 mL), washed with Na$_2$S$_2$O$_5$ solution (20 mL), water (10 mL), brine (20 mL) and dried over anhydrous Na$_2$SO$_4$. The solvents were removed under reduced pressure, and the crude product was purified by column chromatography on silica gel.

2-Morpholino-4*H*-3,1-benzoxazin-4-one (199-morph)[198]

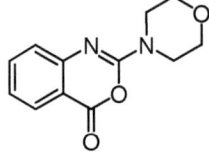

Compound **199**-morph (209 mg, 45%) was obtained following GP10 from *o*-bromophenyl isocyanide (**159**-Br) (364 mg, 2 mmol) and morpholine (174 mg, 2 mmol) and after column chromatography on silica gel (hexane/ethyl acetate 2 : 1, R_f = 0.28) as a colorless solid, m. p. 150–151 °C [lit.[198] 150.5–151.5]. ^{1}H NMR (300 MHz, CDCl$_3$): δ 8.01 (dd, *J* = 8.1, 1.3 Hz, 1 H, Ar-H), 7.66 (ddd, *J* = 8.6, 7.2, 1.9 Hz, 1 H, Ar-H), 7.24 (d, *J* = 8.4 Hz, 1 H, Ar-H), 7.16 (ddd, *J* = 8.1, 7.2, 1.3 Hz, 1 H, Ar-H), 3.80–3.72 ppm (m, 8 H, CH$_2$); ^{13}C NMR (75.5 MHz, CDCl$_3$): δ 159.6 (C), 153.2 (C), 150.4 (C), 136.7 (CH), 128.7 (CH), 124.2 (CH), 123.6 (CH), 112.5 (C), 66.3 (CH$_2$), 44.3 ppm (CH$_2$); MS (EI) *m/z* (%): 232.2 (60) [M$^+$], 146.1 (100); IR (KBr): 2918, 2871, 1768, 1602, 1475, 1308, 1238, 1115, 991, 762, 687 cm^{-1}; elemental analysis calcd (%) for C$_{12}$H$_{12}$N$_2$O$_3$: C 62.06, H 5.21, N 12.06; found: C 61.26, H 5.11, N 12.01.

2-(Aziridin-1-yl)-4*H*-3,1-benzoxazin-4-one (199-azirid)

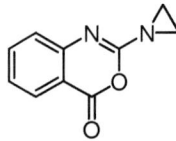

Compound **199**-azirid (188 mg, 50%) was obtained following GP10 from *o*-bromophenyl isocyanide (**159**-Br) (364 mg, 2 mmol) and aziridine (86 mg, 2 mmol) and after column chromatography on silica gel (ethyl acetate, R_f = 0.26) as a colorless solid, m. p. 154–155 °C. ^{1}H NMR (300 MHz, CDCl$_3$): δ 8.15 (dd, *J* = 7.5, 1.3 Hz, 1 H, Ar-H), 7.66 (ddd, *J* = 8.4, 7.2, 1.6 Hz, 1 H, Ar-H), 7.50 (d, *J* = 8.1 Hz, 1 H, Ar-H), 7.32 (ddd, *J* = 8.1, 7.2, 1.3 Hz, 1 H, Ar-H), 4.76 (t, *J* = 8.4 Hz, 2 H, CH$_2$), 4.37 ppm (t, *J* = 8.4 Hz, 2 H, CH$_2$); ^{13}C NMR (75.5 MHz, CDCl$_3$): δ 160.8 (C), 155.4 (C), 148.9 (C), 134.8 (CH), 126.5 (CH), 126.1 (CH), 124.7 (CH), 118.3 (C), 65.8 (CH$_2$), 42.2 ppm (CH$_2$); MS (EI) *m/z* (%): 188.1 (100) [M$^+$], 146.1 (86); IR (KBr): 1696, 1640, 1609, 1562, 1473, 1418, 1263, 1135, 1015, 980, 863, 769, 692 cm^{-1}; elemental analysis calcd (%) for C$_{10}$H$_8$N$_2$O$_2$: C 63.82, H 4.28, N 14.89; found: C 63.62, H 4.21, N 14.67.

2*H*-3,1-Benzoxazine-2,4(1*H*)-dion (isatoic anhydride, 209)[199]

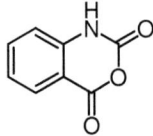

Compound **209** (199 mg, 61%) was obtained employing the same procedure as for **199**-morph, using water (1 mL) instead of morpholine, and after column chromatography on silica gel (hexane/ethyl acetate 1 : 1, R_f = 0.12) as a colorless solid, m. p. 233–234 °C [lit.[199] 233 °C]. ^{1}H NMR (300 MHz, DMSO[d6]): δ 11.68 (br s, 1 H, NH), 7.91 (dd, *J* = 7.5, 1.1 Hz, 1 H, Ar-H), 7.73 (ddd, *J* = 7.9, 7.2, 1.5 Hz, 1 H, Ar-

H), 7.25 (ddd, J = 8.3, 8.3, 1.1 Hz, 1 H, Ar-H), 7.16 ppm (d, J = 8.3 Hz, 1 H, Ar-H); ^{13}C NMR (75.5 MHz, DMSO[d6]): δ 159.7 (C), 147.0 (C), 141.3 (C), 136.8 (CH), 128.8 (CH), 123.4 (CH), 115.2 (CH), 110.1 ppm (C); MS (EI) *m/z* (%): 163.0 (48) [M$^+$], 119.1 (100), 92.1 (54); IR (KBr): 3068 (br) (NH), 1772 (C=O), 1616, 1488, 1364, 1261, 1009, 765, 680, 649 cm^{-1}; elemental analysis calcd (%) for $C_8H_5NO_3$: C 58.90, H 3.09, N 8.59; found: C 58.55, H 3.05, N 8.19.

Experimental Procedures for the Compounds Described in Chapter 4 "Synthesis of 1-Substituted Benzimidazoles from o-Bromophenyl Isocyanide and Amines"

General Procedure for the Synthesis of 1-Substituted Benzimidazoles 232 and 3-Substituted 3*H*-thieno[2,3-d]imidazoles 235 (G11)

In a 10 mL Schlenk flask were placed 2-bromophenyl isocyanide (**159**-Br) (364 mg, 2 mmol) or 2-bromo-3isocyanothiophene (**234**) (376 mg, 2 mmol), cesium carbonate (652 mg, 4 mmol), CuBr (14.4 mg, 5 mol%), 1,10-phenanthroline (36 mg, 10 mol%) and the respective amine (if solid). The flask was sealed with a rubber septum, evacuated and refilled with dry nitrogen three times. Anhydrous degassed DMF (or a solution of a respective liquid amine in DMF) was introduced to the flask from a syringe. The septum was replaced with a glass stopper. The mixture was stirred at r.t. for 2 h, then warmed to 90 °C for ca. 30 min and stirred at this temperature for 14 h. After this time, the mixture was cooled, and the solvent was removed in vacuo. The residue was dissolved in CH_2Cl_2 and water (60 and 15 mL, respectively), the aqueous phase was extracted with CH_2Cl_2 (2×20 mL), and the combined organic phases were washed with brine, dried over Na_2SO_4 and concentrated to give a crude product, which was purified by flash chromatography on silica gel.

2-(1-Methyl-1*H*-indol-3-yl)ethanamine (230d)[200]

A solution of tryptamine (3.2 g, 20 mmol) in anhydrous DMF (40 mL) was added dropwise at r.t. within 20 min to a 60% suspension of sodium hydride in mineral oil (0.88 g, 22 mmol) in anhydrous DMF (60 mL). The mixture was stirred at r.t. for 30 min, cooled to 0 °C, and MeI (3.12 g, 1.37 mL, 22 mmol) was added dropwise. The resulting mixture was stirred at r.t. for 1 h, and the solvent was removed in vacuo. The residue was dissolved in water (300 mL) and extracted with EtOAc (3 × 50 mL). The combined organic phases were dried over Na_2SO_4, and the solvents were removed under reduced pressure to give a crude product, which was purified by column chromatography on silica gel (CH_2Cl_2/MeOH/Et_3N 85 : 10 : 5, R_f = 0.37) to give 2.58 g (74%) of the title compound as a yellow oil. ^{1}H NMR (300 MHz, $CDCl_3$): δ 7.57 (d, J = 7.9 Hz, 1 H, Ar-H), 7.29–7.18 (m, 2 H, Ar-H), 7.09 (t, J = 7.9 Hz, 1 H, Ar-H), 6.87 (s, 1 H, 2-H), 3.71 (s, 3 H, CH_3), 3.00–2.96 (m, 2 H, CH_2), 2.90–2.86 (m, 2 H, CH_2), 2.16 ppm (br s, 2 H, NH_2); ^{13}C NMR (75.5 MHz, $CDCl_3$): δ 137.0 (C), 127.7 (C), 126.8 (CH), 121.5 (CH), 118.8 (CH), 118.6 (CH), 111.9 (C), 109.1

(CH), 42.2 (CH$_3$), 32.5 (CH$_2$), 28.9 ppm (CH$_2$); MS (70 eV, EI) *m/z* (%): 174 (16) [M$^+$], 144 (100); IR (KBr): 3051, 2929. 1615, 1473, 1377, 1327, 1250, 1131, 1012, 741 cm^{-1}.

N-Benzyl-*N*'-(2-bromophenyl)formamidine (231a)

To a stirred solution of 2-bromophenyl isocyanide (**159**-Br) (364 mg, 2 mmol) and benzylamine (**230a**) (214 mg, 2 mmol) in DMF (2 mL) was added CuI (19.1 mg, 0.01 mmol), and the mixture was stirred at r.t. until no more isocyanide was detectable by TLC. The solvent was removed in vacuo, and the product (425 mg, 74%) was isolated by column chromatography on silica gel (CH$_2$Cl$_2$/MeOH 30 : 1, R_f = 0.31) as a yellow solid, m.p. 89–90 °C. This product was identical with an authentic sample isolated from the reaction of 2-bromophenyl isocyanide (**159**-Br) and benzylamine (**230a**) with Et$_3$N as a base (see Table 13, entry 6 of main part). ^{1}H NMR (300 MHz, CDCl$_3$): δ 7.59 (s, 1 H, N=CH), 7.53 (d, *J* = 7.9 Hz, 1 H, Ar-H), 7.45–7.23 (m, 5 H, Ar-H), 7.19 (t, *J* = 7.5 Hz, 1 H, Ar-H), 6.88 (d, *J* = 7.2 Hz, 2 H, Ar-H), 5.04 (br s, 1 H, NH), 4.62 ppm (br s, 2 H, CH$_2$); ^{13}C NMR (75.5 MHz, CDCl$_3$): δ 150.0 (CH), 150.5 (CH), 138.4 (C), 132.7 (CH), 128.5 (2 CH), 128.0 (2 CH), 127.3 (CH), 123.8 (CH), 121.0 (C), 118.5 (C), 44.9 ppm (CH$_2$); MS (ESI) *m/z* (%): 289.0/291.0 (100/95) [M+Na$^+$]; IR (KBr): 3214, 3018, 1698, 1493, 1469, 1449, 1369, 1201, 1024, 751, 719 cm^{-1}; HRMS (ESI) calcd for C$_{14}$H$_{14}$N$_2$Br$^+$ [M+H$^+$]: 289.0335; found: 289.0339.

1-Benzyl-1*H*-benzo[d]imidazole (232a)[201]

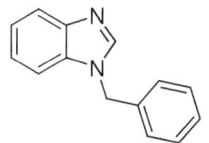

Compound **232a** (283 mg, 68%) was obtained from 2-bromophenyl isocyanide (**159**-Br) (364 mg, 2 mmol) and benzylamine (**230a**) (214 mg, 2 mmol) following the GP11, after column chromatography (CH$_2$Cl$_2$/MeOH 20 : 1, R_f = 0.27) as a colorless solid, m. p. 115–116 °C. [lit. 116–117 °C] ^{1}H NMR (300 MHz, CDCl$_3$): δ 7.93 (s, 1 H, N=CH), 7.83 (d, *J* = 7.2 Hz, 1 H, Ar-H), 7.35–7.21 (m, 6 H, Ar-H), 7.16 (m, 2 H, Ar-H), 5.32 ppm (s, 2 H, CH$_2$); ^{13}C NMR (75.5 MHz, CDCl$_3$): δ 144.2 (C), 135.4 (CH), 128.9 (CH), 128.5 (C), 128.2 (CH), 127.4 (C), 127.0 (CH), 123.0 (CH), 122.0 (CH), 120.4 (CH), 110.0 (CH), 48.7 ppm (CH$_2$); MS (70 eV, EI) *m/z* (%): 208.1 (74) [M$^+$], 91.1 (100); IR (KBr): 3010, 2943, 2154, 1609, 1466, 1184, 1076, 753, 720, 694 cm^{-1}; HRMS (ESI) calcd for C$_{14}$H$_{13}$N$_2$$^+$ [M+H$^+$]: 209.10732; found: 209.10730.

1-*n*-Propyl-1*H*-benzo[d]imidazole (232b)[202]

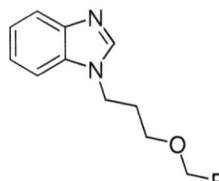

Compound **232b** (209 mg, 65%) was obtained from 2-bromophenyl isocyanide (**159**-Br) (364 mg, 2 mmol) and *n*-propylamine (**230b**) (118 mg, 2 mmol) following the GP11, after column chromatography (CH$_2$Cl$_2$/MeOH 20 : 1, R_f = 0.38) as a yellow oil. ^{1}H NMR (300 MHz, CDCl$_3$): δ 7.87 (s, 1 H, NCH), 7.84–7.78 (m, 1 H, Ar-H), 7.40–7.37 (m, 1 H, Ar-H), 7.32–7.24 (m, 2 H, Ar-H), 4.10 (t, *J* = 7.2 Hz, 2 H, CH$_2$), 1.89 (m, 2 H, CH$_2$), 0.93 ppm (t, *J* = 7.2 Hz, 3 H, CH$_3$); ^{13}C NMR (75.5 MHz, CDCl$_3$, APT): δ 143.8 (C), 142.9 (CH), 122.6 (CH), 121.8 (CH), 120.2 (CH), 109.6 (CH), 46.6 (CH$_2$), 23.0 (CH$_2$), 11.2 ppm (CH$_3$); IR (KBr): 2966, 2933, 2876, 1635, 1496, 1459, 1384, 1367, 1331, 1288, 1259, 1212, 745 cm^{-1}; MS (70 eV, EI) *m/z* (%): 160.0 (58) [M$^+$], 131.0 (100); HRMS (ESI) calcd for C$_{10}$H$_{13}$N$_2$$^+$ [M+H$^+$]: 161.10732; found: 161.10730.

1-(3-(Benzyloxy)propyl)-1*H*-benzo[d]imidazole (232c)

Compound **232c** (350 mg, 66%) was obtained from 2-bromophenyl isocyanide (**159**-Br) (364 mg, 2 mmol) and 3-(benzyloxy)propyl-1-amine (**230c**) (330 mg, 2 mmol) following the GP11, after column chromatography (CH$_2$Cl$_2$/MeOH 20 : 1, R_f = 0.34) as a yellow oil. ^{1}H NMR (300 MHz, CDCl$_3$): δ 7.82–7.79 (m, 2 H), 7.41–7.31 (m, 6 H), 7.29–7.24 (m, 2 H), 4.46 (s, 2 H, PhCH$_2$), 4.29 (t, *J* = 6.8 Hz, 2 H, OCH$_2$), 3.38 (t, *J* = 5.6 Hz, 2 H, NCH$_2$), 2.11 ppm (hept, *J* = 5.6 Hz, 2 H, CH$_2$); ^{13}C NMR (75.5 MHz, CDCl$_3$): δ 143.7 (C), 143.2 (C), 137.8 (C), 128.4 (CH), 127.8 (CH), 127.7 (2 CH), 126.8 (CH), 122.8 (CH), 122.0 (2 CH), 120.2 (CH), 109.6 (CH), 73.1 (CH$_2$), 65.9 (CH$_2$), 41.5 (CH$_2$), 29.8 ppm (CH$_2$); IR (KBr): 3060, 2929, 2861, 1496, 1456, 1366, 1286, 1254, 1201, 1106, 748, 699 cm^{-1}; MS (70 eV, EI) *m/z* (%): 266.2 (16) [M$^+$], 175.1 (16), 160.1 (80), 132.1 (100), 91.0 (52); HRMS (ESI) calcd for C$_{17}$H$_{19}$N$_2$O$^+$ [M+H$^+$]: 267.1492; found: 267.1499.

1-(2-(1-Methyl-1*H*-indol-3-yl)ethyl)-1*H*-benzo[d]imidazole (232d)

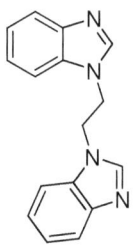

Compound **232d** (322 mg, 59%) was obtained from 2-bromophenyl isocyanide (**159**-Br) (364 mg, 2 mmol) and 2-(1-methyl-1*H*-indol-3-yl)ethylamine (**230d**) (348 mg, 2 mmol) following the GP11, after column chromatography (CH$_2$Cl$_2$/MeOH 30 : 1, R_f = 0.23) as a colorless solid, m. p. 110–111°C. ^{1}H NMR (300 MHz, CDCl$_3$): δ 7.83–7.78 (m, 1 H, Ar-H), 7.60 (s, 1 H, N=CH), 7.56 (d, J = 7.9 Hz, 1 H, Ar-H), 7.43–7.37 (m, 1 H, Ar-H), 7.32–7.23 (m, 4 H, Ar-H), 7.15 (ddd, J = 7.9, 6.8, 1.1 Hz, 1 H, Ar-H), 6.52 (s, 1 H, NCH), 4.44 (t, J = 6.8 Hz, 2 H, CH$_2$), 3.66 (s, 3 H, CH$_3$), 3.28 ppm (t, J = 6.8 Hz, 2 H, CH$_2$); ^{13}C NMR (75.5 MHz, CDCl$_3$, APT): δ 143.9 (C), 143.2 (CH), 137.1 (C), 133.6 (C), 127.3 (CH), 127.2 (C), 122.7 (CH), 121.94 (CH), 121.86 (CH), 120.4 (CH), 119.2 (CH), 118.2 (CH), 109.9 (C), 109.54 (CH), 109.51 (CH), 45.7 (CH$_2$), 32.6 (CH$_3$), 26.0 ppm (CH$_2$); IR 3053, 2923, 1636, 1614, 1493, 1474, 1326, 1287, 1223, 1150, 1126, 1065, 1008, 925, 890, 861, 745, 731 cm^{-1}; MS (ESI) *m/z* (%): 573.3 (8) [2M+Na$^+$], 298.1 (23) [M+Na$^+$], 276.1 (100) [M+H$^+$]; HRMS (ESI) calcd for C$_{18}$H$_{18}$N$_3^+$ [M+H$^+$]: 276.1495; found: 276.1501.

1-(2-(1*H*-Benzo[d]imidazol-1-yl)ethyl)-1*H*-benzo[d]imidazole (232e)[203]

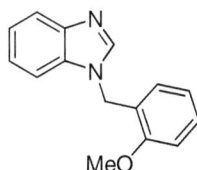

Compound **232e** (220 mg, 42%) was obtained from 2-bromophenyl isocyanide (**159**-Br) (728 mg, 4 mmol) and ethylenediamine (**230e**) (120 mg, 2 mmol) following the GP11, after column chromatography (CH$_2$Cl$_2$/MeOH 10 : 1, R_f = 0.25) as a colorless solid, m. p. 223–224 °C. ^{1}H NMR (300 MHz, DMSO[D$_6$]): δ 7.90 (s, 2 H, NCH), 7.62–7.59 (m, 2 H, Ar-H), 7.43–7.40 (m, 2 H, Ar-H), 7.19–7.14 (m, 4 H, Ar-H), 4.73 ppm (s, 4 H, CH$_2$); ^{13}C NMR (75.5 MHz, DMSO[d6]): δ 143.7 (C), 143.2 (CH), 133.6 (C), 122.2 (CH), 121.5 (CH), 119.3 (CH), 109.9 (CH), 43.8 ppm (CH$_2$); IR (KBr): 3091, 3052, 1609, 1489, 1458, 1361, 1328, 1288, 1261, 1201, 1170, 1149, 1119, 883, 747 cm^{-1}; MS (ESI) *m/z* (%): 547.2 (27) [2M+Na$^+$], 285.1 (57) [M+Na$^+$], 263.1 (100) [M+H$^+$]; HRMS (ESI) calcd for C$_{16}$H$_{15}$N$_4^+$ [M+H$^+$]: 263.1291; found: 263.1290.

1-(2-Methoxybenzyl)-1*H*-benzo[d]imidazole (232f)

Compound **232f** (318 mg, 67%) was obtained from 2-bromophenyl isocyanide (**159**-Br) (364 mg, 2 mmol) and 2-methoxybenzylamine (**230f**) (274 mg, 2 mmol) following the GP11, after column chromatography

(CH$_2$Cl$_2$/MeOH 20 : 1, R_f = 0.33) as a colorless solid, m. p. 75–77 °C. ^{1}H NMR (300 MHz, CDCl$_3$): δ 7.97 (s, 1 H, N=CH), 7.83–7.77 (m, 1 H, Ar-H), 7.41–7.37 (m, 1 H, Ar-H), 7.31–7.22 (m, 3 H, Ar-H), 7.03 (d, J = 7.5, 1.9 Hz, 1 H, Ar-H), 6.88 (m, 2 H, Ar-H), 5.33 (s, 2 H, CH$_2$), 3.85 ppm (s, 3 H, CH$_3$); ^{13}C NMR (75.5 MHz, CDCl$_3$, APT): δ 157.1 (C), 129.7 (CH), 129.0 (CH), 123.7 (C), 123.1 (C), 122.7 (CH), 121.9 (CH), 120.6 (CH), 120.2 (CH), 112.4 (C), 110.5 (CH), 110.0 (C), 55.3 (CH$_3$), 44.2 ppm (CH$_2$); IR (KBr): 3051, 2934, 1600, 1495, 1457, 1286, 1249, 1024, 745 cm^{-1}; MS (70 eV, EI) m/z (%): 238.1 (62) [M$^+$], 121.1 (100), 91.1 (66); HRMS (ESI) calcd for C$_{15}$H$_{15}$N$_2$O$^+$ [M+H$^+$]: 239.1179; found: 239.1169.

1-(3,5-Dimethoxybenzyl)-1H-benzo[d]imidazole (232g)

Compound **232g** (350 mg, 65%) was obtained from 2-bromophenyl isocyanide (**159**-Br) (364 mg, 2 mmol) and 3,5-dimethoxybenzylamine (**230g**) (401 mg, 2 mmol) following the GP11, after column chromatography (CH$_2$Cl$_2$/MeOH 20 : 1, R_f = 0.36) as a yellow oil. ^{1}H NMR (300 MHz, CDCl$_3$): δ 7.95 (s, 1 H, NCH), 7.84–7.81 (m, 1 H), 7.33–7.25 (m, 3 H), 6.43–6.31 (m, 3 H), 5.28 (s, 2 H, CH$_2$), 3.72 ppm (s, 6 H, OCH$_3$); ^{13}C NMR (75.5 MHz, CDCl$_3$, APT): δ 161.3 (C), 144.0 (C), 143.2 (CH), 137.8 (C), 134.0 (C), 123.1 (CH), 122.3 (CH), 120.4 (CH), 110.0 (CH), 105.2 (CH), 99.7 (CH), 55.3 (CH$_3$), 48.9 ppm (CH$_2$); IR (KBr): 1614, 1497, 1459, 1431, 1351, 1290, 1205, 1158, 1066, 832, 745 cm^{-1}; MS (70 eV, EI) m/z (%): 268.2 (40) [M$^+$], 194.1 (100), 151.1 (44), 121.1 (26); HRMS (ESI) calcd for C$_{16}$H$_{17}$N$_2$O$_2^+$ [M+H$^+$]: 269.1285; found: 269.1286.

1-[(Fur-2-yl)methyl]-1H-benzo[d]imidazole (232h)

Compound **232h** (181 mg, 46%) was obtained from 2-bromophenyl isocyanide (**159**-Br) (364 mg, 2 mmol) and furfurylamine (**230h**) (194 mg, 2 mmol) following the GP11, after column chromatography (CH$_2$Cl$_2$/MeOH 20 : 1, R_f = 0.30) as a yellow oil. ^{1}H NMR (300 MHz, CDCl$_3$): δ 7.92 (s, 1 H, N=CH), 7.83–7.76 (m, 1 H, Ar-H), 7.47–7.41 (m, 1 H, Ar-H), 7.37 (t, J = 1.1 Hz, 1 H, furyl-H), 7.33–7.24 (m, 2 H, Ar-H), 6.33 (m, 2 H, furyl-H), 5.28 ppm (s, 2 H, CH$_2$); ^{13}C NMR (75.5 MHz, CDCl$_3$, APT): δ 152.6 (C), 148.4 (C), 143.1 (CH), 123.1 (C), 123.1 (CH), 122.2 (CH), 120.4 (CH), 110.6 (CH), 110.3 (CH), 109.7 (CH), 109.1 (CH), 41.7 ppm (CH$_2$); IR (KBr): 1615, 1495, 1459, 1364, 1287, 1270, 1238, 1200, 1167, 1147, 1012, 885, 746 cm^{-1}; MS (70 eV, EI) m/z (%): 198.0 (38) [M$^+$], 81 (100), 53 (26); HRMS (ESI) calcd for C$_{12}$H$_{11}$N$_2$O$^+$ [M+H$^+$]: 199.08659; found: 199.08658.

1-[4-(Trifluoromethyl)benzyl]-1*H*-benzo[d]imidazole (232i)

Compound **232i** (302 mg, 55%) was obtained from 2-bromophenyl isocyanide (**159**-Br) (364 mg, 2 mmol) and 4-(trifluoromethyl)benzylamine (**230i**) (350 mg, 2 mmol) following the GP11, after column chromatography (CH$_2$Cl$_2$/MeOH 20 : 1, R_f = 0.34) as a colorless solid, m. p. 75–76 °C. ^{1}H NMR (300 MHz, CDCl$_3$): δ 7.96 (s, 1 H, N=CH), 7.84 (d, J = 6.8 Hz, 1 H, Ar-H), 7.58 (d, J = 7.9 Hz, 2 H, Ar-H), 7.33–7.21 (m, 5 H, Ar-H), 5.41 ppm (s, 2 H, CH$_2$); ^{13}C NMR (75.5 MHz, CDCl$_3$, APT): δ 139.5 (q, J = 1.1 Hz, C), 130.8 (C), 130.3 (C), 127.5 (CH), 127.1 (CH), 126.0 (q, J = 3.9 Hz, CH), 125.6 (C), 123.4 (CH), 122.5 (CH), 122.0 (CH), 120.6 (CH), 109.8 (CH), 48.2 ppm (CH$_2$); IR (KBr): 1617, 1496, 1420, 1326, 1162, 1109, 1066, 1015, 826, 745 cm^{-1}; MS (70 eV, EI) *m/z* (%): 276.2 (100) [M$^+$], 159.0 (89); HRMS (ESI) calcd for C$_{15}$H$_{12}$N$_2$F$_3^+$ [M+H$^+$]: 277.09471; found: 277.09482.

1-Cyclopropyl-1*H*-benzo[d]imidazole (232j)[204]

Compound **232j** (125 mg, 40%) was obtained from 2-bromophenyl isocyanide (**159**-Br) (364 mg, 2 mmol) and cyclopropylamine (**230j**) (114 mg, 2 mmol) following the GP11, after column chromatography (CH$_2$Cl$_2$/MeOH 20 : 1, R_f = 0.26) as a yellow oil. ^{1}H NMR (300 MHz, CDCl$_3$): δ 7.91 (s, 1 H, NCH), 7.79–7.75 (m, 1 H, Ar-H), 7.57–7.53 (m, 1 H, Ar-H), 7.33–7.24 (m, 2 H, Ar-H), 3.38–3.31 (m, 1 H, cPr-CH), 1.16–1.08 (m, 2 H, cPr-CH$_2$), 1.04–0.99 ppm (m, 2 H, cPr-CH); ^{13}C NMR (75.5 MHz, CDCl$_3$): δ 143.6 (C), 143.3 (CH), 135.0 (C), 122.9 (CH), 122.2 (CH), 120.2 (CH), 110.2 (CH), 25.2 (CH), 5.6 ppm (CH$_2$); IR (KBr): 3094, 1643, 1615, 1494, 1460, 1315, 1289, 1239, 1031, 746 cm^{-1}; MS (70 eV, EI) *m/z* (%): 158.0 (74) [M$^+$], 157 (100), 131 (47); HRMS (ESI) calcd for C$_{10}$H$_{11}$N$_2^+$ [M+H$^+$]: 159.09167; found: 159.09171.

1-Cyclohexyl-1*H*-benzo[d]imidazole (232k)[205]

Compound **232k** (184 mg, 46%) was obtained from 2-bromophenyl isocyanide (**159**-Br) (364 mg, 2 mmol) and cyclohexylamine (**230k**) 198 mg, 2 mmol) following the GP11, after column chromatography (CH$_2$Cl$_2$/MeOH 40 : 1, R_f = 0.22) as a yellowish solid, m.p. 72–74 °C [lit. 74–75 °C]. ^{1}H NMR (300 MHz, CDCl$_3$): δ 7.98 (s, 1 H, N=CH), 7.83–7.78 (m, 1 H, Ar-H), 7.45–7.40 (m, 1 H,

Ar-H), 7.30–7.23 (m, 2 H, Ar-H), 4.18 (s, 1 H, CH), 2.20 (d, J = 11.3 Hz, 2 H, CH$_2$), 1.96 (d, J = 13.2 Hz, 2 H, CH$_2$), 1.85–1.72 (m, 3 H, CH$_2$), 1.57–1.42 (m, 2 H, CH$_2$), 1.38–1.27 ppm (m, 1 H, CH$_2$); ^{13}C NMR (75.5 MHz, CDCl$_3$, APT): δ 143.8 (C), 140.3 (CH), 133.3 (C), 122.4 (CH), 121.9 (CH), 120.3 (CH), 110.0 (CH), 55.3 (CH), 33.2 (CH$_2$), 25.6 (CH$_2$), 25.3 ppm (CH$_2$); IR (KBr): 3109, 3052, 2933, 2855, 1634, 1490, 1456, 1287, 1216, 889, 744 cm^{-1}; MS (70 eV, EI) m/z (%): 200 (100) [M$^+$], 157 (27), 118 (66); HRMS (ESI) calcd for C$_{13}$H$_{17}$N$_2^+$ [M+H$^+$]: 201.13862; found: 201.13863.

1-p-Tolyl-1*H*-benzo[d]imidazole (232l)[206]

Compound **232l** (172 mg, 41%) was obtained from 2-bromophenyl isocyanide (**159**-Br) (364 mg, 2 mmol) and *p*-toluidine (**230l**) (214 mg, 2 mmol) following the GP11, after column chromatography (CH$_2$Cl$_2$/MeOH 40 : 1, R_f = 0.35) as a colorless solid, m. p. 51–52 °C [lit.[206] 50–54]. ^{1}H NMR (300 MHz, CDCl$_3$): δ 8.08 (s, 1 H, NCH), 7.89–7.86 (m, 1 H, Ar-H), 7.52–7.49 (m, 1 H, Ar-H), 7.39–7.30 (m, 6 H, Ar-H), 2.44 ppm (s, 3 H, CH$_3$); ^{13}C NMR (75.5 MHz, CDCl$_3$, APT): δ 143.9 (C), 138.0 (C), 134.1 (CH), 133.8 (C), 133.7 (C), 130.5 (CH), 123.9 (CH), 123.5 (CH), 122.6 (CH), 120.4 (CH), 110.4 (CH), 21.0 ppm (CH$_3$); IR (KBr): 3055, 3062, 1692, 1611, 1518, 1490, 1456, 1289, 1231, 1205, 822, 744 cm^{-1}; MS (70 eV, EI) m/z (%): 208.0 (100) [M$^+$]; HRMS (ESI) calcd for C$_{14}$H$_{13}$N$_2^+$ [M+H$^+$]: 209.10732; found: 209.10734.

1-(2-Bromophenyl)-1*H*-benzo[d]imidazole (232n)

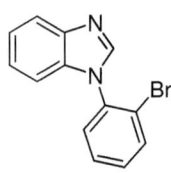

The compound **232n** (105 mg, 38%) was obtained from 2-bromophenyl isocyanide (**159**-Br) (364 mg, 2 mmol) and *tert*-butylamine (**230m**) (146 mg, 2 mmol) following GP11, after column chromatography (CH$_2$Cl$_2$/MeOH 30 : 1, R_f = 0.32) as a red oil. An authentic sample of benzimidazole **232n** prepared from **159**-Br and *o*-bromoaniline (**230n**) in 42% yield, was identical with the previous one. ^{1}H NMR (300 MHz, CDCl$_3$): δ 8.04 (s, 1 H, NCH), 7.90 (dd, J = 6.4, 1.5 Hz, 1 H), 7.81 (dd, J = 7.9, 1.1 Hz, 1 H), 7.53–7.26 (m, 5 H), 7.19 ppm (dd, J = 6.8, 1.9 Hz, 1 H); ^{13}C NMR (75.5 MHz, CDCl$_3$, APT): δ 143.1 (C), 142.9 (CH), 135.1 (C), 134.2 (C), 134.1 (CH), 130.5 (CH), 129.0 (CH), 128.6 (CH), 123.6 (CH), 122.7 (CH), 121.4 (CH), 120.4 (CH), 110.5 ppm (CH); IR (KBr): 1613, 1586, 1494, 1454, 1306, 1288, 1230, 1203, 1056, 1030, 786, 744, 721 cm^{-1}; MS (ESI) m/z (%): 567.0 (2M+Na$^+$), 273.0 (100) [M+H$^+$]; HRMS (ESI) calcd for C$_{13}$H$_{10}$BrN$_2^+$ [M+H$^+$]: 273.0022; found: 273.0030.

3-Benzyl-3H-thieno[2,3-d]imidazole (235a)

Compound **235a** (210 mg, 49%) was obtained from 2-bromo-3isocyanothiophene (**234**) (376 mg, 2 mmol) and benzylamine (**230a**) (214 mg, 2 mmol) following the GP11, after column chromatography (CH_2Cl_2/MeOH 40 : 1, R_f = 0.30) as a colorless solid, m. p. 102–103 °C. ^{1}H NMR (300 MHz, $CDCl_3$): δ 7.72 (s, 1 H, NCH), 7.36 (m, 3 H, Ph), 7.27 (m, 2 H, Ph), 7.12 (d, J = 5.3, 1 H, thienyl-H), 6.92 (d, J = 5.3, 1 H, thienyl-H), 5.19 ppm (s, 2 H, CH_2); ^{13}C NMR (75.5 MHz, $CDCl_3$, APT): δ 148.7 (C), 141.8 (CH), 134.1 (C), 131.6 (C), 129.0 (CH), 128.7 (CH), 128.1 (CH), 120.7 (CH), 116.6 (CH), 51.2 ppm (CH_2); IR (KBr): 1635, 1516, 1456, 1436, 1392, 1354, 1252, 1188, 1092, 1035, 907, 734 cm^{-1}; MS (EI) m/z (%): 214.2 (44) [M$^+$], 91.1 (100); HRMS (ESI) calcd for $C_{12}H_{11}N_2S$ [M+H$^+$]: 215.06375; found: 215.06369.

3-(3-(Benzyloxy)propyl)-3H-thieno[2,3-d]imidazole (235c)

Compound **235c** (242 mg, 44%) was obtained from 2-bromo-3-isocyanothiophene (**234**) (376 mg, 2 mmol) and 3-(benzyloxy)prop-1-yl-amine (**230c**) (330 mg, 2 mmol) following the GP11, after column chromatography (CH_2Cl_2/MeOH 40 : 1, R_f = 0.23) as a yellow oil. ^{1}H NMR (300 MHz, $CDCl_3$): δ 7.55 (s, 1 H, NCH), 7.39–7.28 (m, 5 H, Ph), 7.14 (d, J = 5.3, 1 H, thienyl-H), 6.98 (d, J = 5.3, 1 H, thienyl-H), 4.48 (s, 2 H, OCH_2), 4.21 (t, J = 6.8 Hz, 2 H, CH_2), 3.42 (t, J = 6.0 Hz, 2 H, CH_2), 2.14 ppm (pent, J = 6.0 Hz, 2 H, CH_2); ^{13}C NMR (75.5 MHz, $CDCl_3$, APT): δ 148.5 (C), 142.2 (CH), 137.9 (C), 131.4 (C), 128.4 (CH), 127.7 (CH), 127.6 (CH), 120.2 (CH), 116.8 (CH), 73.1 (CH_2), 65.8 (CH_2), 43.9 (CH_2), 29.3 ppm (CH_2); IR (KBr): 2926, 2862, 1634, 1517, 1454, 1391, 1366, 1251, 1195, 1098, 733 cm^{-1}; MS (EI) m/z (%): 272.3 (28) [M$^+$], 206.1 (100) 91.1 (94); HRMS (ESI) calcd for $C_{15}H_{17}N_2OS^+$ [M+H$^+$]: 273.10561; found: 273.10560.

3-(2-(1-Methyl-1H-indol-3-yl)ethyl)-3H-thieno[2,3-d]imidazole (235d)

Compound **235d** (250 mg, 44%) was obtained from 2-bromo-3-isocyanothiophene (**234**) (376 mg, 2 mmol) and 2-(1-methyl-1H-indol-3-yl)ethanamine (**230d**) (348 mg, 2 mmol) following the GP11, after column chromatography (CH_2Cl_2/MeOH 40 : 1, R_f = 0.23) as a yellow oil. ^{1}H NMR (300 MHz, $CDCl_3$): δ 7.54 (d, J = 7.5, 1 H), 7.41 (s, 1 H, NCH), 7.31–7.22 (m, 2 H), 7.16–7.11 (m, 2 H), 6.98 (dd, J = 5.3, 1.1 Hz, 1 H, thienyl-H), 6.59 (s, 1 H, indolyl-2H), 4.33 (t, J = 7.2 Hz, 2 H, CH_2), 3.67 (s, 3 H, CH_3), 3.31 ppm (t, J = 7.2 Hz, 2 H, CH_2); ^{13}C NMR (75.5 MHz, $CDCl_3$, APT): δ 148.5 (C), 142.2 (CH), 137.0 (C), 131.3 (C), 127.8 (C), 127.2 (CH), 121.8 (CH),

120.2 (CH), 119.1 (CH), 118.2 (CH), 116.8 (CH), 109.6 (C), 109.5 (CH), 48.0 (CH$_2$), 32.6 (CH$_3$), 25.4 ppm (CH$_2$); IR (KBr): 3443, 1640, 1517, 1474, 1435, 1380, 1328, 1250, 1201, 903, 738 cm^{-1}; MS (ESI) *m/z* (%): 585.2 (33) [2M+Na$^+$], 304.1 (100) [M+Na$^+$]; HRMS (ESI) calcd for C$_{16}$H$_{16}$N$_3$S$^+$ [M+H$^+$]: 282.1059; found: 282.1061

D. Summary and Outlook

A variety of transformations, which isocyanides can undergo *en route* to different *N*-heterocycles is almost as diverse and versatile, as organic chemistry itself. The examples shown in the Introduction of this thesis covered only cases, in which both C and N atoms of the isocyano group are integrated into newly formed *N*-heterocycles. In the major part of such catalyzed or base-induced processes two possible routes are realized: 1) an initial deprotonation of the isocyanide is followed by its addition and (or) cyclization or 2) an addition to the isocyano group (or its insertion) is followed by a cyclization of the thus formed reactive intermediate. Base-induced anionic cyclizations are supplemented with some radical processes and transition metal-catalyzed (mediated) reactions as well as organocatalytic transformations. Some shown cyclizations have been found to proceed with high stereo- and enantioselectivities. The versatility and simplicity of such processes has found its reflection in syntheses of various natural products themelves as well as key precursors. Although much of isocyanide chemistry and particularly the syntheses of *N*-heterocycles, have been explored in last 30–40 years, we were convinced even before starting this work, that many methods still remained uncovered. Thus, the main objective of this doctoral thesis has been to find and explore new approaches to *N*-heterocycles from isocyanides.

Two different new syntheses of substituted pyrroles from isocyanides and acetylenes have been developed (see Chapter 1). The formal cycloaddition of α-metallated methyl isocyanides **63** onto the triple bond of electron-deficient acetylenes **64** reported recently by de Meijere and Larionov represents a direct and convenient approach to 2,3,4-trisubstituted pyrroles **65**. The scope and limitations of this reaction were further elaborated in this study. Some new alkyl-, aryl- and hetarylpropiolates **168** were employed in the efficient synthesis of 2,3,4-trisubstituted pyrroles **173** (7 examples, 68–94%). The terminal acceptor-substituted acetylenes, such as methyl propiolate (**168h**) have been shown to provide the corresponding 2,4-disubstituted pyrroles in their reaction with substituted methyl isocyanides **63**, albeit in lower yields (4 examples, 7–44%). Some test experiments towards optimization of the reaction conditions (solvent, temperature, catalyst) are presented herein along with the description of a plausible mechanism. Next, we tried to fathom the possibility of employing unactivated acetylenes in their reaction with **63**. Thus, a novel Cu(I)-mediated synthesis of 2,3-disubstituted pyrroles **178** by reaction of copper acetylides derived from unactivated terminal alkynes **167** with substituted methyl isocyanides **63** has been developed. After the optimization of reaction conditions, 11 examples of such 2,3-disubstituted pyrroles **178** have been synthesized (5–88% yield). The proposed mechanism of this new transformation was confirmed by some additional experiments.

Metallated isocyanides, as mentioned above, may be versatile precursors for various *N*-heterocycles. We envisaged, that *ortho*-metallated phenyl isocyanide **188**-Li and related compounds (**200**) which have not been known before, might be versatile intermediates for the synthesis of particular heterocycles as well. In Chapter 2, the generation and further reactions of *ortho*-lithiophenyl isocyanide **188**-Li, the first example of a ring- metallated aryl isocyanide known so far, are described. Thus, **188**-Li conveniently obtained by halogen–lithium exchange on *ortho*-bromophenyl isocyanide (**159**-Br), was trapped with various electrophiles to provide corresponding 2-substituted phenyl isocyanides **192** (5 examples, 55–88% yield). In the reaction of **188**-Li with dimethylformamide, 2-(formylamino)-benzaldehyde **196** was formed unexpectedly and isolated in 76% yield. The latter presumably arose by hydrolysis of the initially formed 1,3-benzoxazine derivative **194**. The reactions of **188**-Li with isocyanates and isothiocyanates afforded, after treatment of the reaction mixture with water, pharmaceutically relevant 3-substituted 3*H*-quinazoline-4-ones and 3*H*-quinazolin-4-thiones **191** (9 examples, 69–91% yield). Treatment of the same mixtures after the reaction of *ortho*-lithiophenyl isocyanide **188**-Li with an isocyanate containing lithiated intermediates **190**-Li with a second electrophile, provided the corresponding 2,3-disubstituted 3*H*-quinazoline-4-ones **191** (6 examples, 54–85% yield). In two cases, this trapping could proceed intramolecularly as an appropriate functional group was provided in the isocyanates themselves. Thus, the naturally occurring alkaloids deoxyvasicinone (**191n**) and tryptanthrine (**191o**) were synthesized in 72 and 85% yield, respectively, in a one-pot procedure following this strategy.

In Chapter 3, the reactions of *ortho*-lithiophenyl isocyanide (**188**-Li) and some of its heteroanalogues (3-isocyano-2-thienyllithium **216** and 3-isocyano-2-pyridyllithium **218**) with aldehydes, ketones and carbon dioxide are considered in detail. *ortho*-Lithiophenyl (-hetaryl) isocyanides of the general type **200** react at −78 °C with aldehydes to provide the corresponding isocyanobenzylalcohols **204** (36–89%, 9 examples), and with ketones to form the respective 4*H*-3,1-benzoxazines **201** (48–78%, 3 examples), when the mixture was treated with water before work-up. Treatment of the same mixtures at −78 °C with other electrophiles provided in moderate to good yields 2-substituted 4*H*-3,1-benzoxazines **201l**-R, **206**, **207** and in one case, the mixed carbonate **205** of the isocyanoalcohol **204d**. 2-Lithiated 4*H*-3,1-benzoxazines of type **198** (and their heteroanalogues generated from lithiated isocyanides of type **200**) have been shown to undergo two types of unprecedented rearrangements providing isobenzofuran-1(3*H*)-imines (iminophthalanes) **210** (and its heteroanalogues **211**, **219l**) or indolin-2-ones **215** (and its heteroanalogue **217k**), depending on the reaction conditions and substitution patterns. Proposed mechanisms of these novel rearrangements include pericyclic ring opening in **198** with destruction of benzene ring aromatic character followed by two types of recyclizations to give *N*-metallated oxoindoles **215** or isobenzofuran-1(3*H*)-imines **210**. Isocyanoalcohols **204** in turn were converted to 4*H*-3,1-benzoxazines **201** or isobenzofuran-1(3*H*)-imines **210** (or its heteroanalogue **211d**) under Cu(I) catalysis (66–86%, 8 examples). 4*H*-3,1-Benzoxazin-4-ones **199**-Nu and isatoic anhydride **209** were obtained by the reaction of **188**-Li with carbon dioxide followed by trapping of the lithiated intermediate with iodine and subsequent reactions with nucleophiles (45–60%, 3 examples).

Transition-metal catalyzed processes have become one of the most important parts of modern organic chemistry in general and particularly in the synthesis of heterocycles. Therefore, the exploration of new such processes employing isocyanides is strongly required. In the last part of this thesis (Chapter 4), a novel copper-catalyzed synthesis of 1-substituted benzimidazoles **232** from *o*-bromoaryl isocyanide (**159**-Br) and primary amines (**230**) is presented. The optimization of the reaction conditions revealed that the best yields of benzimidazoles are achieved, when the reaction is performed in DMF with Cs_2CO_3 as a base and CuBr/1,10-Phenanthroline as a catalyst. Importantly, the temperature of the reaction mixture should be gradually increased up to 90 °C to achieve highest yields. Under optimized conditions, **159**-Br reacts with various primary amines in the presence of the Cu(I) catalyst to afford 1-substituted benzimidazoles **232** in moderate to good yields (38–70%, 13 examples). Analogously, 2-bromo-3-isocyanothiophene (**234**) furnishes 3-substituted 3*H*-thieno[2,3-d]imidazoles **235** (44–49%, 3 examples). Mechanistically, 1-substituted benzimidazoles **232** and their heteroanalogues **235** are believed to result from a sequential reaction consisting of a copper-catalyzed addition of an amine **230** onto an isocyano group of **159**-Br followed by a copper-catalyzed intramolecular *N*-arylation of the thus formed formamidine (**231**). Interestingly, the Cu(I)-catalyzed reaction of **159**-Br with *tert*-butylamine did not provide the corresponding *N*-*tert*-butyl benzimidazoles (**232m**), but gave 1-(2-bromophenyl)benzimidazole **232n** in 38% yield. The supposed rationale for this fact involves the in situ formation of 2-bromoaniline (**230n**) and subsequent reaction of **159**-Br with it. This assumption was confirmed by independent reaction of **159**-Br with **230n**, which also provided **232n** in 42% yield.

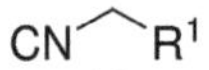

63
CN–R¹

R¹ = CO₂Me
CO₂Et
SO₂Tol
Ph
$pC_6H_4NO_2$

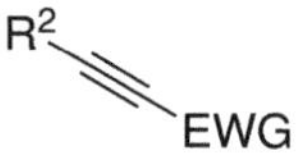

64
R²–≡–EWG

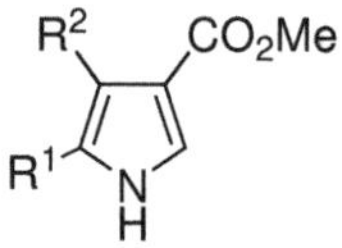

168
R²–≡–CO₂Me

R² = cPr
CH(OMe)Me
$pEtOC_6H_4$
pFC_6H_4
$pCF_3C_6H_4$
2-pyridyl
2-thienyl
H

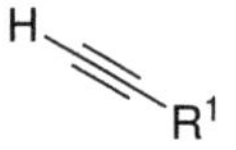

173

R¹ = CO₂Me, R² = cPr (**173aa**)
R¹ = CO₂Et, R² = cPr (**173ba**)
R¹ = CO₂Me, R² = CH(OMe)Me (**173ab**)
R¹ = CO₂Me, R² = $pEtOC_6H_4$ (**173ac**)
R¹ = CO₂Me, R² = pFC_6H_4 (**173ad**)
R¹ = CO₂Me, R² = $pCF_3C_6H_4$ (**173ae**)
R¹ = CO₂Me, R² = 2-pyridyl (**173af**)
R¹ = CO₂Me, R² = 2-thienyl (**173ag**)
R¹ = CO₂Et, R² = H (**173bh**)
R¹ = SO₂Tol, R² = H (**173ch**)
R¹ = Ph, R² = H (**173eh**)
R¹ = $pC_6H_4NO_2$, R² = H (**173fh**)

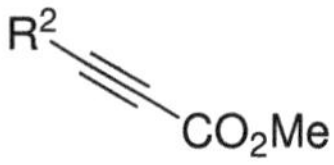

167
H–≡–R¹

R¹ = nBu
CH₂OMe
CH(OMe)Me
Ph
cPr
tBu
2-pyridyl
secBu
(CH₂)₂OH

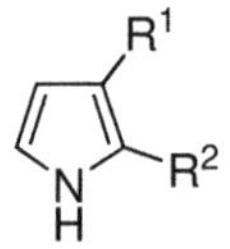

178

R¹ = nBu, R² = CO₂Et (**178ba**)
R¹ = CH₂OMe, R² = CO₂Et (**178bb**)
R¹ = CH(OMe)Me, R² = CO₂Et (**178bc**)
R¹ = Ph, R² = CO₂Et (**178bd**)
R¹ = cPr, R² = CO₂Et (**178be**)
R¹ = tBu, R² = CO₂Et (**178bf**)
R¹ = 2-pyridyl, R² = CO₂Et (**178bg**)
R¹ = secBu, R² = CO₂Et (**178bh**)
R¹ = nBu, R² = CO₂Et (**178bi**)
R¹ = nBu, R² = CO₂tBu (**178ca**)
R¹ = nBu, R² = $pC_6H_4NO_2$ (**178bi**)

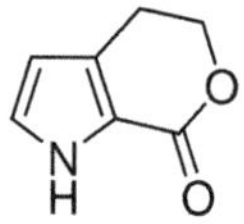

179

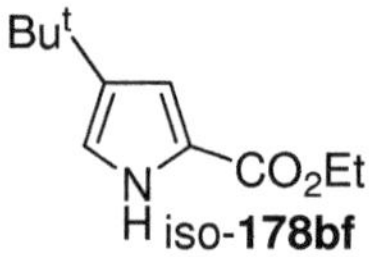

iso-**178bf** (Buᵗ substituted pyrrole with CO₂Et)

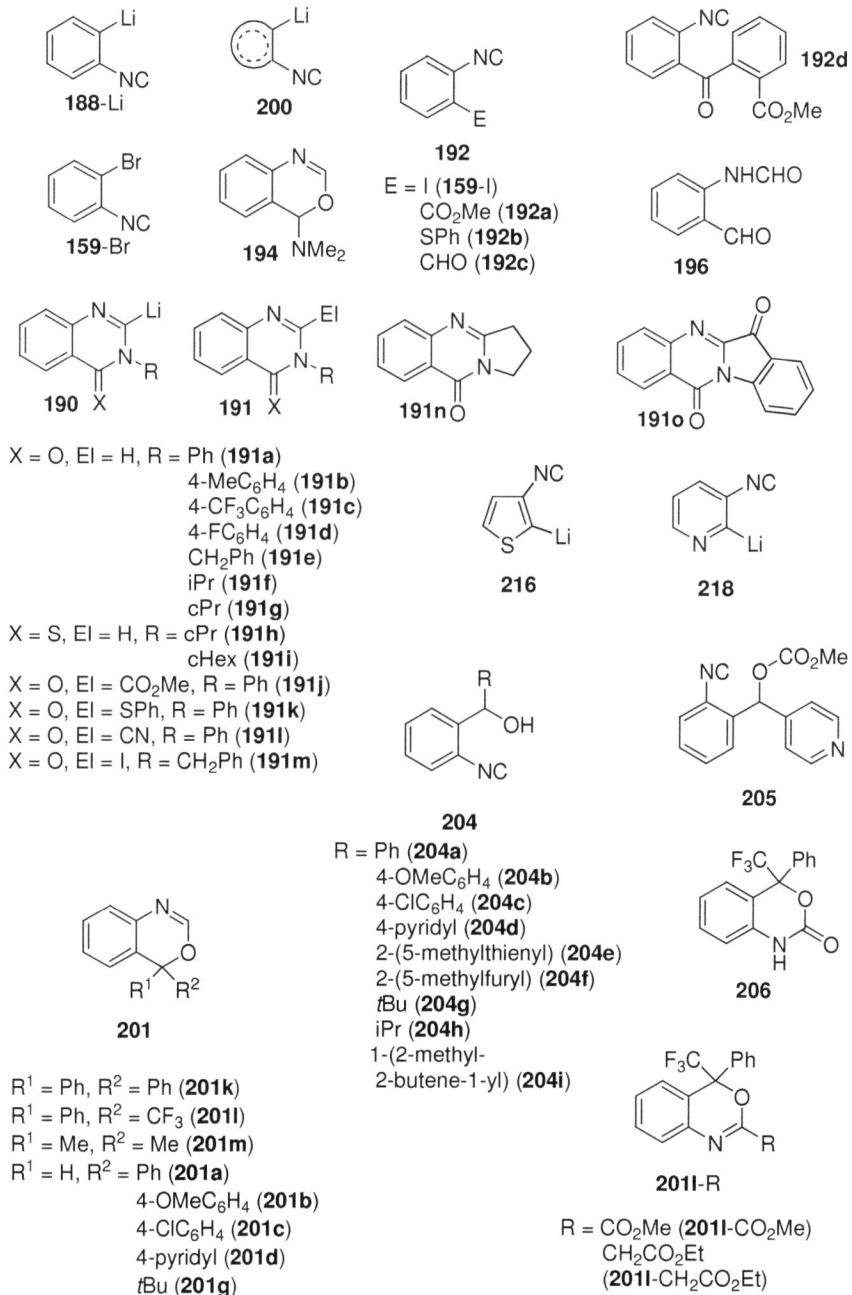

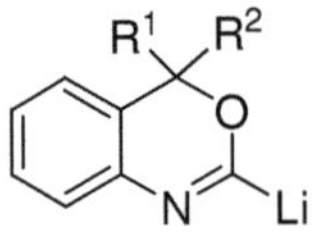

198

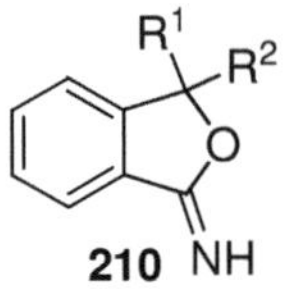

210 R¹ = H, R² = 2-(5-methyl-furyl) (**210f**)
R¹ = H, R² = iPr (**210h**)
R¹ = Me, R² = CF₃ (**210o**)

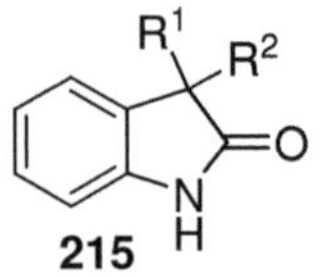

215 R¹ = H, R² = 2-pyridyl (**215n**)
R¹ = Ph, R² = Ph (**215k**)

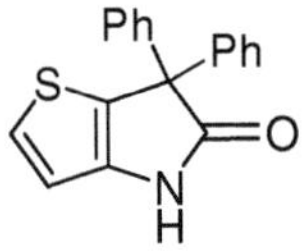

217k

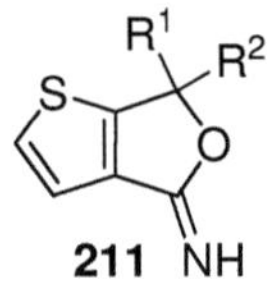

211 R¹ = H, R² = 4-pyridyl (**211d**)
R¹ = Ph, R² = CF₃ (**211l**)

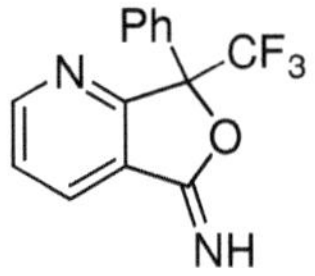

219l

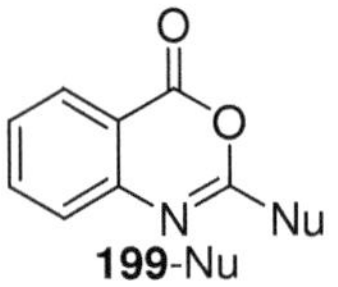

199-Nu
Nu = N-morpholine (**199-morph**)
Nu = N-aziridine (**199-azirid**)

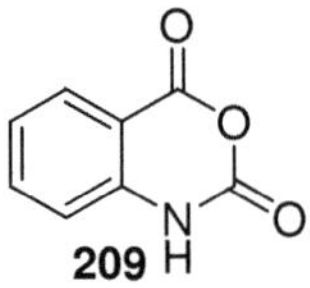

209

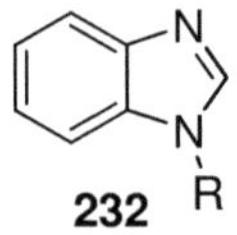

232
R = Ph (**232a**)
nPr (**232b**)
(CH₂)₃OCH₂Ph (**232c**)
(CH₂)₂(3-N-methylindolyl) (**232d**)
CH₂(2-MeOC₆H₄) (**232f**)
CH₂(3,5-MeOC₆H₃) (**232g**)
CH₂(2-furyl) (**232h**)
CH₂(4-CF₃C₆H₄) (**232i**)
cPr (**232j**)
cHex (**232k**)
4-MeC₆H₄ (**232l**)
tBu (**232m**)
2-BrC₆H₄ (**232n**)

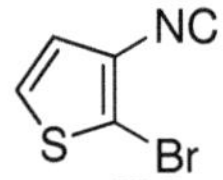

230 RNH₂

234

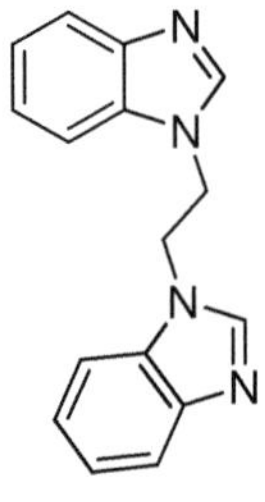

232e

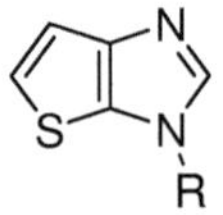

235
R = CH₂Ph (**235a**)
(CH₂)₃OCH₂Ph (**235c**)
(CH₂)₂(3-N-methylindolyl) (**232d**)

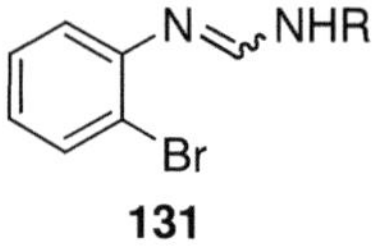

131

E. References and Comments

[1] a) A. Gautier, *Ann. Chem. Pharm.* **1868**, *146*, 119–124; Isocyanides as new isomers of cyanides are mentioned and announced previously (but not described) in: A. Gautier, *Ann. Chem. Pharm.* **1867**, *142*, 289–294.

[2] A. W. Hofmann, *Ann. Chem. Pharm.* **1867**, *144*, 114–120.

[3] a) I. Ugi, R. Meyr, *Angew. Chem.* **1958**, *70*, 702–703; b) I. Ugi, U. Fetzer, U. Eholzer, H. Knupfer, K. Offermann, *Angew. Chem.* **1965**, *77*, 492–504; *Angew. Chem. Int. Ed. Engl.* **1965**, *4*, 472–484.

[4] a) W. P. Weber, G. W. Gokel, Tetrahedron Lett. 1972, 1637–1640; b) W. P. Weber, G. W. Gokel, I. K. Ugi, *Angew. Chem.* **1972**, *84*, 587; *Angew. Chem. Int. Ed. Engl.* **1972**, *11*, 530–531.

[5] For a general review, see: a) M. Suginome, Y. Ito, In *Science of Synthesis* Vol. 19 (Ed.: S.-I. Murahashi), Thieme, Stuttgart, **2004**, pp. 445–530, and references cited therein.

[6] For reviews, see: a) A. Dömling, I. Ugi, *Angew. Chem.* **2000**, *112*, 3300–3344; *Angew. Chem. Int. Ed.* **2000**, *39*, 3168–3210; b) H. Bienayme, C. Hulme, G. Oddon, P. Schmitt, *Chem. Eur. J.* **2000**, *6*, 3321–3329; c) J. Zhu, *Eur. J. Org. Chem.* **2003**, 1133–1144; d) V. Nair, C. Rajesh, A. U. Vinod, S. Bindu. A. R. Sreekanth, J. S. Mathen, L. Balagopal, *Acc. Chem. Res.* **2003**, *36*, 899–907; d) A. Dömling, *Chem. Rev.* **2006**, *106*, 17–89; e) L. El Kaim, L. Grimaud, *Tetrahedron* **2009**, *65*, 2153–2171.

[7] For some representative examples, see: a) N. Chatani, T. Hanafusa, *J. Org. Chem.* **1991**, *56*, 2166–2170; b) E. Kroke, S. Willms, M. Weidenbruch, W. Saak, S. Pohl, H. Marsmann, *Tetrahedron Lett.* **1996**, *37*, 3675–3678; c) S. Kamijo, Y. Yamamoto, *J. Am. Chem. Soc.* **2002**, *124*, 11940–11945; d) N. Chatani, M. Oshita, M. Tobisu, Y. Ishii, S. Murai, *J. Am. Chem. Soc.* **2003**, *125*, 7812–7813; e) G. Bez, C.-G. Zhao, *Org. Lett.* **2003**, *5*, 4991–4993; f) M. Oshita, K. Yamashita, M. Tobisu, N. Chatani, *J. Am. Chem. Soc.* **2005**, *127*, 761–766; g) P. Fontaine, G. Masson, J. Zhu, *Org. Lett.* **2009**, *11*, 1555–1558.

[8] a) R. F. Heck, In: *Palladium Reagents in Organic Synthesis*; Academic Press: New York, 1985. b) J. Tsuji, In: *Palladium Reagents and Catalysts*; John Wiley: Chichester, U.K., 1995. c) Y. Ito, M. Suginome, In: *Handbook of Organopalladium Chemistry for Organic Synthesis* (Eds.: Negishi, E.; de Meijere, A.), Wiley, New York 2002.

[9] For reviews on metal-isocyanide complexes, see: a) Y. Yamamoto, H. Yamazaki, *Coord. Chem. Rev.* **1972**, *8*, 225–239; b) P. M. Treichel, *Adv. Organomet. Chem.* **1973**, *11*, 21–86; c) E. Shingleton, H. E. Oosthuizen, *Adv. Organomet. Chem.* **1983**, *22*, 209–310.

[10] M. Suginome, Y. Ito, *Adv. Polym. Sci.* **2004**, *171* (polymer synthesis) 77–136.

[11] a) T. Fukuyama, X. Chen, G. Peng, *J. Am. Chem. Soc.* **1994**, *116*, 3127–3128; b) Y. Kobayashi, T. Fukuyama, *J. Heterocycl. Chem.* **1998**, *35*, 1043–1055; c) H. Tokuyama, Y. Kaburagi, X. Chen, T. Fukuyama, *Synthesis* **2000**, 429–434; For a review, see d) H. Tokuyama, T. Fukuyama, *Chem. Rec.* **2002**, *2*, 37–45.

[12] H. Josien, S.-B. Ko, D. Born, D. P. Curran, *Chem. Eur. J.* **1998**, *4*, 67–83:

[13] a) For a review on tandem radical reactions with isocyanides, see: I. Ryu, N. Sonoda, D. P. Curran, *Chem. Rev.* **1996**, *96*, 177–194.

[14] U. Schöllkopf, F. Gerhart, *Angew. Chem.* **1968**, *80*, 842–843; *Angew. Chem. Int. Ed. Engl.* **1968**, *7*, 805–806.

[15] For reviews, see: a) D. Hoppe, *Angew. Chem.* **1974**, *86*, 878–893; b) U. Schöllkopf, *Angew.Chem.* **1977**, *89*, 351–360; *Angew. Chem. Int. Ed. Engl.* **1977**, *16*, 339–348; c) U Schöllkopf, *Pure Appl. Chem.* **1979**, *51*, 1347–1355; d) K. Matsumoto, T. Moriya, M. Suzuki, *J. Synth. Org. Chem., Jpn.* **1985**, *43*, 764–776.

[16] a) D. H. R. Barton, S. Z. Zard, *J. Chem. Soc., Chem. Commun.* **1985**, 1098–1100; b) D. H. R. Barton, J. Kervagoret, S. Z. Zard, *Tetrahedron* **1990**, *46*, 7587–7598; c) J. L. Sessler, A. Mozattari, M. Johnson, *Org. Synth.* **1992**, *70*, 68–77; Coll. Vol. 9 **1998**, 242–251.

[17] T. D. Lash, J. R. Belletini, J. A. Bastian, K. B. Couch, *Synthesis* **1994**, 170–172.

[18] J. Tang, J. G. Verkade, *J. Org. Chem.* **1994**, *59*, 7793–7802.

[19] A. Bhattacharya, S. Cherukuri, R. E. Plata, N. Patel, V. Tamez, Jr., J. A. Grosso, M. Peddicordb, V. A. Palaniswam, *Tetrahedron Lett.* **2006**, *47*, 5481–5484.

[20] a) N. Ono, H. Hironaga, K. Ono, S. Kaneko, T. Murashima, T. Ueda, C. Tsukamura, T. Ogawa, *J. Chem. Soc., Perkin Trans. 1* **1996**, 417–423; b) T D. Lash, P. Chandrasekar, A. T. Osuma, S. T. Chaney, J. D. Spence, *J. Org. Chem.*, **1998**, *63*, 8455–8469.

[21] a) P. Magnus, P. Halazy, *Tetrahedron Lett.* **1984**, *25*, 1421–1424; b) G. Haake, D. Struve, F.-P. Montforts, *Tetrahedron Lett.* **1994**, *35*, 9703–9704; c) D. P. Arnold, L. Burgess-Dean, J. Hubbard, M. A. Rahman, *Aust. J. Chem.* **1994**, *47*, 969–974; d) Y. Abel, F.-P. Montforts, *Tetrahedron Lett.* **1997**, *38*, 1745–1748; e) W. Schmidt, F.-P. Montforts, *Synlett* **1997**, 903–904; f) S. Ito, T. Murashima, N. Ono, *J. Chem. Soc., Perkin Trans. 1* **1997**, 3161–3165; g) Y. Abel, E. Haake, G. Haake, W. Schmidt, D. Struve, A. Walter, F.-P. Montforts, *Helv. Chim. Acta* **1998**, *81*, 1978–1996; h) H. Uno, M. Tanaka, T. Inoue, N. Ono, *Synthesis*, **1999**, *3*, 471–474.

[22] a) W. Huebsch, R. Angerbauer, P. Fey, H. Bischoff, D. Petzinna, D. Schmidt, G. Thomas, *Eur. Pat. Appl.*; Bayer, A.-G.; Fed. Rep. Ger.: Ep, 1989; p 36; b) J. L. Bullington, R. R. Wolff, P. F. Jackson, *J. Org. Chem.*, **2002**, *67*, 9439–9442.

[23] N. C. Misra, K. Panda, H. Ila, H. Junjappa, *J. Org. Chem.* **2007**, *72*, 1246–1251.

[24] Y. Fumoto, T. Eguchi, H. Uno, N. Ono, *J. Org. Chem.*, **1999**, *64*, 6518–6521.

[25] U. Robben, I. Lindner, W. Gärtner, *J. Am. Chem. Soc.* **2008**, *130*, 11303-11311.

[26] N. Ono, H. Kawamura, M. Bougauchi, K. Maruyama, *Tetrahedron*, **1990**, *46*, 7483–7496.

[27] A. M. van Leusen, G. J. M. Boerma, R. B. Helmholdt, H. Siderius, J. Strating, *Tetrahedron Lett.* **1972**, *23*, 2367–2368. For Reviews, see: d) Review: D. van Leusen, A. M. van Leusen, *Org. React.* **2001**, *57*, 417–666; e) V. K. Tandon, S. Rai, *Sulfur Rep.* **2003**, *24*, 307–385.

[28] For reviews, see: a) A. M. van Leusen, D. van Leusen In *Encyclopedia for Organic Synthesis*; L. A. Paquette Ed.; Wiley: New York, 1995, Vol. 7, pp 4973–4979; b) A. M. van Leusen, *Lect. Heterocycl. Chem.* **1980**, *5*, S111–S122;

[29] a) A. M. van Leusen, B. E. Hoogenboom, H. Siderius, *Tetrahedron Lett.* **1972**, *13*, 2369–2372; b) B. A. Kulkarni, A. Ganesan, *Tetrahedron Lett.* **1999**, *40*, 5637–5638.

[30] a) A. M. van Leusen, J. Wildeman, O. Oldenziel, *J. Org. Chem.* **1977**, *42*, 1153–1159; b) R. ten Have, M. Huisman, A. Meetsma, A. M. van Leusen, *Tetrahedron* **1997**, *53*, 11355–11368.

[31] For synthesis of imidazoles fused to other heterocyclic systems with TosMIC, see: a) P. Chen, J. C. Barrish, E. Iwanowicz, J. Lin, M. S. Bednarz, B.-C. Chen, *Tetrahedron Lett.* **2001**, *42*, 4293–4295; b) B.-C. Chen, R. Zhao, M. S. Bednarz, B. Wang, J. E. Sundeen, J. C. Barrish, *J. Org. Chem.* **2004**, *69*, 977–979.

[32] b) A. M. van Leusen, H. Siderius, B. E. Hoogenboom, D. van Leusen, *Tetrahedron Lett.* **1972**, *13*, 5337–5340; c) D. van Leusen E. Flentge, A. M. van Leusen, *Tetrahedron* **1991**, *47*, 4639–4644.

[33] H. P. Dijkstra, R. ten Have, A. M. van Leusen, *J. Org. Chem.* **1998**, *63*, 5332–5338.

[34] N. D. Smith, D. Huang, N. D. P. Cosford, *Org. Lett.* **2002**, *4*, 3537–3539.

[35] a) J. Moskal, R. van Stralen, D. Postma, A. M. van Leusen, *Tetrahedron Lett.* **1986**, *27*, 2173–2176; b) J. Moskal, A. M. van Leusen, *J. Org. Chem.* **1986**, *51*, 4131–4139.

[36] a) A. R. Katritzky, Y. X. Chen, K. Yannakopoulou, P. Lue, *Tetrahedron Lett.* **1989**, *30*, 6657–6660; b) A. R. Katritzky, D. Cheng, R. P. Musgrave, *Heterocycles* **1997**, *44*, 67–70.

[37] a) T. Saegusa, Y. Ito, H. Kinoshita, S. Tomita, *J. Org. Chem.* **1971**, *36*, 3316–3323; b) Y. Ito, T. Matsuura, T. Saegusa, *Tetrahedron Lett.* **1985**, *26*, 5781–5784;

[38] T. Hayashi, E. Kishi, V. Soloshonok, Y. Uozumi, *Tetrahedron Lett.* **1996**, *37*, 4969–4972.
[39] R. Grigg, M. I. Lansdell, M. Thornton-Pett, *Tetrahedron*, **1999**, *55*, 2025–2044.
[40] B. Trost, *Science* **1991**, *254*, 1471–1477.
[41] a) Y. Ito, M. Sawamura, T. Hayashi, *J. Am. Chem. Soc.* **1986**, *108*, 6405–6406; b) Y. Ito, M. Sawamura, M. Kobayashi, T. Hayashi, *Tetrahedron Lett.* **1987**, *28*, 6215–6218; c) Y. Ito, M. Sawamura, E. Shirakawa, K. Hayashizaki, T. Hayashi, *Tetrahedron Lett.* **1988**, *29*, 235–238; d) Y. Ito, M. Sawamura, E. Shirakawa, K. Hayashizaki, T. Hayashi, *Tetrahedron* **1988**, *44*, 5253–5262; e) Y. Ito, M. Sawamura, T. Hayashi, *Tetrahedron Lett.* **1988**, *29*, 239–240; f) Y. Ito, M. Sawamura, H. Hamashima, T. Emura, T. Hayashi, *Tetrahedron Lett.* **1989**, *30*, 4681–4684; e) T. Hayashi, M. Sawamura, Y. Ito, *Tetrahedron* **1992**, *48*, 1999–2012; For a concise review, see: E. M. Carreira, A. Fetters, C. Marti, *Org. React.* (Hoboken, NY) **2006**, *67*, 1–216.
[42] a) Y. Ito, M. Sawamura, T. Hayashi, *Tetrahedron Lett.* **1988**, *29*, 6321–6324; b) M. Sawamura, Y. Ito, T. Hayashi, *Tetrahedron Lett.* **1989**, *30*, 2247–2250; c) Sawamura, Y. Ito, T. Hayashi, *J. Org. Chem.* **1990**, *55*, 5935–5936; d) M. Sawamura, Y. Nakayama, T. Kato, Y. Ito, *J. Org. Chem.* **1995**, *60*, 1727–1732.
[43] a) S. D. Pastor, A. Togni, *J. Am. Chem. Soc.* **1989**, *111*, 2333–2334; b) A. Togni, R. Häusel, *Synlett*, **1990**, 633-; c) A. Togni, S. D. Pastor, G. Rihs, *J. Organomet. Chem.* **1990**, *381*, C21-; d) A. Togni, S. D. Pastor, *J. Org. Chem.* **1990**, *55*, 1649–1664.
[44] a) R. Nesper, P. S. Pregosin, K. Püntener, M. Wörle, *Helv. Chim. Acta* **1993**, *76*, 2239–2249; b) F. Gorla, A. Togni, L. M. Venanzi, A. Albinati, F. Lianza, *Organometallics* **1994**, *13*, 1607–1616; c) J. M. Longmire, X. Zhang, M. Shang, *Organometallics* **1998**, *17*, 4374–4379; c) Y. Motoyama, H. Kawakami, K. Shimozono, K. Aoki, H. Nishiyama, *Organometallics*, **2002**, *21*, 3408–3416.
[45] a) X.-T. Zhou, Y.-R. Lin, L.-X. Dai, J. Sun, L.-J. Xia, M.-H. Tang, *J. Org. Chem.* **1999**, *64*, 1331–1334; b) X.-T. Zhou, Y.-R. Lin, L.-X. Dai, *Tetrahedron Asymm.* **1999**, *10*, 855–862.
[46] Y.-R. Lin, X.-T. Zhou, L.-X. Dai, *J. Org. Chem.* **1997**, *62*, 1799–1803.
[47] J. Audin, K. S. Kumar, L. Eriksson, K. J. Szabo, *Adv. Synth. Cat.* **2007**, *349*, 2585–2594.
[48] J. Aydin, A. Ryden, K. J. Szabo, *Tetrahedron Assym.* **2008**, *19*, 1867–1870.
[49] D. Benito-Garagorri, V. Bocokic, K. Kirchner, *Tetrahedron Lett.* **2006**, *47*, 8641–8644.
[50] a) S. Kamijo, C. Kanazawa, Y. Yamamoto, *J. Am. Chem. Soc.* **2005**, *127*, 9260–9266; b) S. Kamijo, C. Kanazawa, Y. Yamamoto, *Tetrahedron Lett.* **2005**, *46*, 2563–2566.

[51] O. V. Larionov, A. de Meijere, *Angew. Chem.* **2005**, *117*, 5809–5813; *Angew. Chem. Int. Ed.* **2005**, *44*, 5664–5667

[52] D. Gao, H. Zhai, M. Parvez, T. G. Back, *J. Org. Chem.* **2008**, *73*, 8057–8068.

[53] C. Kanazawa, S. Kamijo, Y. Yamamoto, *J. Am. Chem. Soc.* **2006**, *128*, 10662–10663.

[54] H. Takaya, S. Kojima, S.-I. Murahashi, *Org. Lett.* **2001**, *3*, 421–424.

[55] U. Schöllkopf, F. Gerhart, R. Schröder, *Angew. Chem.* **1969**, *81*, 701; *Angew. Chem. Int. Ed. Engl.* **1969**, *8*, 672.

[56] a) Y. Ito, K. Kobayashi, T. Saegusa, *J. Am. Chem. Soc.* **1977**, *99*, 3532–3534; b) Y. Ito, K. Kobayashi, N. Seko, T. Saegusa, *Bull. Chem. Soc. Jpn.* **1984**, *57*, 73–84.

[57] Y. Ito, Y. Inubushi, T. Sugaya, K. Kobayashi, T. Saegusa, *Bull. Soc. Chem. Jpn.* **1978**, *51*, 1186–1188.

[58] Ito, Y. Kobayashi, K.; Saegusa, T. *J. Org. Chem.* **1979**, 44, 2030–2032.

[59] Y. Ito, T. Konoike, T. Saegusa, *J. Organomet. Chem.* **1975**, *85*, 395–401.

[60] Y. Ito, K. Kobayashi, T. Saegusa, *Tetrahedron Lett.* **1978**, 2087–2090.

[61] Y. Ito, K. Kobayashi, T. Saegusa, *Tetrahedron Lett.* **1979**, 1039–1042.

[62] Y. Ito, K. Kobayashi, M. Maeno, T. Saegusa, *Chem. Lett.* **1980**, 487–490.

[63] Y. Ito, K. Kobayashi, T. Saegusa, *Chem. Lett.* **1980**, 1563–1566.

[64] a) W. D. Jones, W. P. Kosar, *J. Am. Chem. Soc.* **1986**, *108*, 5640–5641; b) G. C. Hsu, W. P. Kosar, W. D. Jones *Organometallics* **1994**, *13*, 385–396.

[65] K. Kobayashi, T. Nakashima, M. Mano, O. Morikawa, H. Konishi, *Chem. Lett.* **2001**, 602–603.

[66] K. Kobayashi, K. Yoneda, T. Mizumoto, H. Umakoshi, O. Morikawa, H. Konishi, *Tetrahedron Lett.* **2003**, *44*, 4733–4736.

[67] a) H. M. Walborsky, G. E. Niznik, *J. Am. Chem. Soc.* **1969**, *91*, 7778;

[68] G. E. Niznik, W. H. Morrison III, H. M. Walborsky, *J. Org. Chem.* **1974**, *39*, 600–604.

[69] a) H. M. Walborsky, P. Ronman, *J. Org. Chem.* **1978**, *43*, 731–734; b) J. Heinicke, *J. Organomet. Chem.* **1989**, *364*, C17–C21.

[70] A. Orita, M. Fukudome, K. Ohe, S. Murai, *J. Org. Chem.* **1994**, *59*, 477–481.

[71] M. Suginome, T. Fukuda, Y. Ito, *Org. Lett.* **1999**, *1*, 1977–1979.

[72] Y. Ito, E. Ihara, M. Hirai, H. Ohsaki, A. Ohnishi, M. Murakami, *J. Chem. Soc., Chem. Commun.* **1990**, 403–405.

[73] a) K. Kobayashi, K. Yoneda, M. Mano, O. Morikawa, H. Konishi, *Chem. Lett.* **2003**, *32*, 76–77; b) K. Kobayashi, K. Yoneda, K. Miyamoto, O. Morikawa, H. Konishi, *Tetrahedron* **2004**, *60*, 11639–11645.

[74] J. Ichikawa, Y. Wada, H. Miyazaki, T. Mori, H. Kuroki, *Org. Lett.* **2003**, *5*, 1455–1458.

[75] J. Ichikawa, T. Mori, H. Miyazaki, Y. Wada, *Synlett* **2004**, 1219–1222.

[76] a) M. Westling, T. Livinghouse, *Tetrahedron Lett.* **1985**, *26*, 5389–5392; b) M. Westling, R. Smith, T. Livinghouse, *J. Org. Chem.* **1986**, *51*, 1159–1165.

[77] G. Luedtke, M. Westling, T. Livinghouse, *Tetrahedron*, **1992**, *48*, 2209–2222.

[78] G. Luedtke, T. Livinghouse, *J. Chem. Soc., Perkin Trans. 1*, **1995**, 2369–2371.

[79] C. H. Lee, M. Westling, T. Livinghouse, A. C. Williams, *J. Am. Chem. Soc.* **1992**, *114*, 4089–4095.

[80] M. Westling, T. Livinghouse, *J. Am. Chem. Soc.* **1987**, *109*, 590–592.

[81] a) D. J. Hughes, T. Livinghouse, *J. Chem. Soc., Perkin Trans. 1* **1995**, 2373–2374; b) T. Kercher, T. Livinghouse, *J. Org. Chem.* **1997**, *62*, 805–812.

[82] R. Bossio, S. Marcaccini, R. Pepino, *Heterocycles*, **1986**, *24*, 2003–2005.

[83] R. Bossio, S. Marcaccini, R. Pepino, C. Polo, G. Valle, *Synthesis*, **1989**, 641–643.

[84] R. Bossio, S. Marcaccini, R. Pepino, *Heterocycles*, **1986**, *24*, 2411–2413.

[85] For proposed by authors mechanism of this transformation, see: R. Bossio, S. Marcaccini, R. Pepino, T. Torroba, G. Valle, *Synthesis*, **1987**, 1138–1139.

[86] E. Bulka, K. D. Ahlers, E. Tucek, *Chem. Ber.* **1967**, *100*, 1367–1372.

[87] a) N. Sonoda, G. Yamamoto, S. Tsutsumi, *Bull. Soc. Chem. Jpn.* **1972**, *45*, 2937–2938; b) S. Fujiwara, T. Matsuya, H. Maeda, T. Shin-ike, N. Kambe, N. Sonoda, *Synlett* **1999**, 75–76.

[88] S. Fujiwara, Y. Asanuma, T. Shin-ike, N. Kambe, *J. Org. Chem.* **2007**, *72*, 8087–8090.

[89] L. L. Joyce, G. Evindar, R. A. Batey, *Chem. Commun.* **2004**, 446–447.

[90] H. Maeda, T. Matsuya, N. Kambe, N. Sonoda, S. Fujiwara, T. Shin-ike, *Tetrahedron* **1997**, *53*, 12159–12166.

[91] A. V. Lygin, O. V. Larionov, V. S. Korotkov, A. de Meijere *Chem. Eur. J.* **2009**, *15*, 227–236.

[92] a) B. D. Roth, C. J. Blankley, A. W. Chucholowski, E. Ferguson, M. L. Hoefle, D. F. Ortwine, R. S. Newton, C. S. Sekerke, D. R. Slikovic, C. D. Stratton, M. W. Wilson, *J. Med. Chem.* **1991**, *34*, 357–366; b) J. M. Gottesfeld, L. Neely, J. W. Trauger, E. E. Baird, P. B. Dervan, *Nature* **1997**, *387*, 202–205; c) M. Adamczyk, D. D. Johnson, R. E. Reddy, *Angew. Chem.* **1999**, *111*, 3751–3753; *Angew. Chem. Int. Ed.* **1999**, *38*, 3537–3539; d) S.

Depraetere, M. Smet, W. Dehaen, *Angew. Chem.* **1999**, *111*, 3556–3558; *Angew. Chem. Int. Ed.* **1999**, *38*, 3359–3361; e) D. E. N. Jacquot, M. Zöllinger, T. Lindel, *Angew. Chem.* **2005**, *117*, 2336–2338; *Angew. Chem. Int. Ed.* **2005**, *44*, 2295–2298; f) T. Lindel, M. Hochgürtel, M. Assmann, M. Köck, *J. Nat. Prod.* **2000**, *63*, 1566–1569; g) G. Dannhardt, W. Kiefer, *Arch. Pharm.* **2001**, *334*, 183–188; h) D. Seidel, V. Lynch, J. L. Sessler, *Angew. Chem.* **2002**, *114*, 1480–1483; *Angew. Chem. Int. Ed.* **2002**, *41*, 1422–1425; i) J. A. Johnson, N. Li, D. Sames, *J. Am. Chem. Soc.* **2002**, *124*, 6900–6903; j) H. Hoffmann, T. Lindel, *Synthesis* **2003**, 1753–1783; k) H. Garrido-Hernandez, M. Nakadai, M. Vimolratana, Q. Li, T. Doundoulakis, P. G. Harran, *Angew. Chem.* **2005**, *117*, 775–779; *Angew. Chem. Int. Ed.* **2005**, *44*, 765–769.

[93] a) K. Yamaji, M. Masubuchi, F. Kawahara, Y. Nakamura, A. Nishio, S. Matsukuma, M. Fujimori, N. Nakada, J. Watanabe, T. Kamiyama, *J. Antibiot.* **1997**, *50*, 402–411; b) B. Fournier, D. C. Hooper, *Antimicrob. Agents & Chemother.* **1998**, *42*, 121–128; c) D. Perrin, B. van Hille, J.-M. Barret, A. Kruczynski, C. Etievant, B. Imbert, B. T. Hill, *Biochem. Pharmacol.* **2000**, *59*, 807–819; d) F. Micheli, R. Di Fabio, R. Benedetti, A. M. Capelli, P. Cavallini, P. Cavanni, S. Davalli, D. Donati, A. Feriani, S. Gehanne, M. Hamdan, M. Maffeis, F. M. Sabbatini, M. E. Tranquillini, M. V. A. Viziano, *Farmaco* **2004**, 175–183.

[94] a) H. Miyaji, W. Sato, J. L. Sessler, *Angew. Chem.* **2000**, *112*, 1847–1850; *Angew. Chem. Int. Ed.* **2000**, *39*, 1777–1780; b) F.-P. Montforts, O. Kutzki, *Angew. Chem.* **2000**, *112*, 612–614; *Angew. Chem. Int. Ed.* **2000**, *39*, 599–601; c) D. W. Yoon, H. Hwang, C.-H. Lee, *Angew. Chem.* **2002**, *114*, 1835–1837; *Angew. Chem. Int. Ed.* **2002**, *41*, 1757–1759; d) J. O. Jeppesen, J. Becher, *Eur. J. Org. Chem.* **2003**, 3245–3266.

[95] a) V. F. Ferreira, M. C. B. V. de Souza, A. C. Cunha, L. O. R. Pereira, M. L. G. Ferreira, *Org. Prep. Proced. Int.* **2001**, *33*, 411–454; b) D. X. Zeng, Y. Chen, *Synlett* **2006**, 490–492, and references therein; c) M. R. Tracey, R. P. Husung, R. H. Lambeth, *Synthesis* **2004**, 918–922, and references therein; for reviews see: d) T. L. Gilchrist, *J. Chem. Soc., Perkin Trans.* **1999**, *1*, 2849–2866; e) *Chemistry of Heterocyclic Compounds: Pyrroles* (Ed.: R. A. Jones) Wiley: New York, 1990; Vol. 48; f) D. S. Black in *Science of Synthesis, Vol. 5* (Ed.: G. Maas) Thieme, Stuttgart, **2001**, pp. 441–552.

[96] For some recent reports on the synthesis of oligosubstituted pyrroles, see: a) D. J. S. Cyr, N. Martin, B. A. Arndtsen, *Org. Lett.* **2007**, *9*, 449–452; b) B.C. Milgram, K. Eskildsen, S. M. Richter, W. R. Scheidt, K. A. Scheidt, *J. Org.Chem.* **2007**, *72*, 3941–3944; c) D. J. S. Cyr, N. Martin, B. A. Arndtsen, *Org. Lett.* **2007**, *9*, 449–452; d) R. M. Rodriguez, S. L. Buchwald,

Org. Lett. **2007**, *9*, 973–976; e) M. Shindo, Y. Yoshimura, M. Hayashi, H. Soejima, T. Yoshikawa, K. Matsumoto, K. Shishido, *Org. Lett.* **2007**, *9*, 1963–1966; f) S. Su, J. A. Porco, Jr., *J. Am. Chem. Soc.* **2007**, *129*, 7744–7745; g) F. M. Istrate, F. Gagosz, *Org. Lett.* **2007**, *9*, 3181–3184; h) R. Martin, C. H. Larsen, A. Cuenca, S. L. Buchwald, *Org. Lett.* **2007**, *9*, 3379–3382; i) H. Dong, M. Shen, J. E. Redford, B. J. Stokes, A. L. Pumphrey, T. G. Driver, *Org. Lett.* **2007**, *9*, 5191–5194; j) S. Chiba, Y.-F. Wang, G. Lapointe, K. Narasaka, *Org. Lett.* **2008**, *10*, 313–316; k) A. S. Dudnik, A. W. Sromek, M. Rubina, J. T. Kim, A. V. Kel'in, V. Gevorgyan, *J. Am. Chem. Soc.* **2008**, *130*, 1440–1452; l) Y. Lu, X. Fu, H. Chen, X. Du, X. Jia, Y. Liu, *Adv. Synth. Catal.* **2009**, *351*, 129–134; m) P. Fontaine, G. Masson, Y. Zhu, *Org. Lett.* **2009**, *74*, 1555–1558; n) L. Ackermann, R. Sandmann, L. T. Kaspar, *Org. Lett.* **2009**, *11*, 2031–2034.

[97] Only pyrroles prepared by the author are included in this table.

[98] a) Y. Matsuya, K. Hayashi, H. Nemoto, *J. Am. Chem. Soc.* **2003**, *125*, 646–647; b) E. Winterfeldt, *Chem. Ber.* **1964**, *97*, 1952–1958; c) A.W. McCulloch, A.G. McInnes, *Can. J. Chem.* **1974**, *52*, 3569–3576.

[99] A. Skatteboel, *Acta Chem. Scand.* **1959**, *13*, 191–198.

[100] a) H. C. Kolb, M. G. Finn, K. B. Sharpless, *Angew. Chem.* **2001**, *113*, 2056–2075; *Angew. Chem. Int. Ed.* **2001**, *40*, 2004–2021; b) V. V. Rostovtsev, L. G. Green, V. V. Fokin, K. B. Sharpless, *Angew. Chem.* **2002**, *114*, 2708–2711; *Angew. Chem. Int. Ed.* **2002**, *41*, 2596–2599; c) C. M. Tornoe, C. Christensen, M. J. Meldal, *J. Org. Chem.* **2002**, *67*, 3057–3064.

[101] a) G. Wilkinson, T. S. Piper, *J. Inorg. Nucl. Chem.* **1956**, *2*, 32–34; b) F. A. Cotton, T. J. Marks, *J. Am. Chem. Soc.* **1970**, *92*, 5114–5117.

[102] D. B. Beach, F. K. LeGoues, C.-K. Hu, *Chem. Mater.* **1990**, *2*, 216–219.

[103] T. Saegusa, Y. Ito, S. Tomita, *J. Am. Chem. Soc.* **1971**, *93*, 5656–5661.

[104] T. Tsuda, H. Habu, S. Horiguchi, T. Saegusa, *J. Am. Chem. Soc.* **1974**, *96*, 5930–5931.

[105] a) T. Saegusa, Y. Ito, S. Tomita, H. Kinoshita, *J. Org. Chem.* **1970**, *88*, 670–675; b) T. Saegusa, Y. Ito. H. Kinoshita, S. Tomita, *Bull. Chem. Soc. Jap.* **1970**, *48*, 877–879; c) T. Saegusa, I. Murase, Y. Ito, *J. Org. Chem.* **1971**, *36*, 2876–2880; d) T. Saegusa, Y. Ito, S. Tomita, H. Kinoshita, *Bull. Chem. Soc. Jap.* **1972**, *45*, 496–499.

[106] T. Saegusa, K. Yonezawa, I. Murase, T. Konoike, S. Tomita, Y. Ito, *J. Org. Chem.* **1973**, *38*, 2319–2328.

[107] P. C. J. Kamer, M. C. Cleij, R. J. M. Nolte, T. Harada, A. M. F. Hezemans, W. Drenth, *J. Am. Chem. Soc.* **1988**, *110*, 1581–1587.

[108] a) M. Komatsu, Y. Yoshida, M. Uesaka, Y. Ohshiro, T. Agawa, *J. Org. Chem.* **1984**, *49*, 1300–1302; b) V. Amarnath, D. C. Anthony, K. Amarnath, W. M. Valentine, L. A. Wetterau, D. G. Graham, *J. Org. Chem.* **1991**, *56*, 6924–6931; c) G. Dana, O. Convert, J.-P. Girault, E. M. Mulliez, *Can. J. Chem.* **1976**, *54*, 1827–1835; d) H. O. Bayer, *Chem. Ber.* **1970**, *103*, 2356–2367.

[109] For a review, see: J. F. Normant, A. Alexakis, *Synthesis* **1981**, 841–870.

[110] W. J. Gensler, A. P. Mahadevan, *J. Org. Chem.* **1956**, 180–182.

[111] For some recently published transformations of pyrroles, see: direct C-arylation of free (NH)-pyrroles: a) X. Wang, B. S. Lane, D. Sames, *J. Am. Chem. Soc.* **2005**, *127*, 4996–4997; b) R. D. Rieth, N. P. Mankad, E. Calimano, J. P. Sadighi, *Org. Lett.* **2004**, *6*, 3981–3983; addition of pyrroles to unfunctionalized enediynes: c) A. Odedra, C.-J. Wu, T. B. Pratap, C.-W. Huang, Y.-F. Ran, R.-S. Liu, *J. Am. Chem. Soc.* **2005**, *127*, 3406–3412; oxidative cyanation of pyrroles: d) T. Dohi, K. Morimoto, Y. Kiyono, H. Tohma, Y. Kita, *Org. Lett.* **2005**, *7*, 537–540; [4+3] cycloadditions onto pyrroles: e) R. P. Reddy, L. Davies, *J. Am. Chem. Soc.* **2007**, *129*, 10312–10313; f) asymmetric hydrogenation of pyrroles: R. Kuwano, M. Kashiwabara, M. Ohsumi, H. Kusano, *J. Am. Chem. Soc.* **2008**, *130*, 808–809.

[112] a) K. Sakai, M. Suzuki, K.-i. Nunami, N. Yoneda, Y. Onoda, Y. Iwasawa, *Chem. Pharm. Bull.* **1980**, *28*, 2384–2393; b) M. Bergauer, P. Gmeiner, *Synthesis* **2001**, *15*, 2281–2288.

[113] A. V. Lygin, A. de Meijere *Org. Lett.* **2009**, *11*, 389–392.

[114] S. Sinha, M. Srivastava, *Prog. Drug Res.* **1994**, *43*, 143.

[115] T. Nagase, T. Mizutani, S. Ishikawa, E. Sekino, T. Sasaki, T. Fujimura, S. Ito, Y. Mitobe, Y. Miyamoto, R. Yoshimoto, T. Tanaka, A. Ishihara, N. Takenaga, S. Tokita, T. Fukami, N. Sato, *J. Med. Chem.* **2008**, 51, 4780–4789.

[116] S. E. de Laszlo, C. S. Quagliato, W. J. Greenlee, A. A. Patchett, R. S. L. Chang, V. J. Lotti, T.-B. Chen, S. A. Scheck, K. A. Faust, S. S. Kivlighn, T. S. Schorn, G. J. Zingaro, P. K. S. Siegl, *J. Med. Chem.* **1993**, *36*, 3207–3210.

[117] a) N. J. Liverton, D. J. Armstrong, D. A. Claremon, D. C. Remy, J. J. Baldwin, R. J. Lynch, G. Zhang, R. J. Gould, *Bioorg. Med. Chem. Lett.* **1998**, *8*, 483–486; b) W. Zhang, J. P. Mayer, S. E. Hall, J. A. Weigel, *J. Comb. Chem.* **2001**, *3*, 255–256.

[118] A. Gopalsamy, H. Yang, *J. Comb. Chem.* **2000**, *2*, 378–381.

[119] B. E. Evans, K. E. Rittle, M. G. Bock, R. M. DiPardo, R. M. Freidinger, W. L. Whitter, G. F. Lundell, D. F. Veber, P. S. Anderson, R. S. L. Chang, V. J. Lotti, D. J. Cerino, T. B.

Chen, P. J. Kling, K. A. Kunkel, J. P. Springer, J. Hirshfield, *J. Med. Chem.* **1988**, *31*, 2235–2246.

[120] D. A. Horton, G. T. Bourne, M. L. Smythe, *Chem. Rev.* **2003**, *103*, 893–930.

[121] a) B. B. Snider, X. Wu, *Org. Lett.* **2007**, *9*, 4913–4915; b) C. H. Oh, C. H. Song, *Synth. Commun.* **2007**, *37*, 3311–3317; c) M. R. Linder, A. R. Heckeroth, M. Najdrowski, A. Daugschies, D. Schollmeyer, C. Miculka, *Bioorg. Med. Chem. Lett.* **2007**, *17*, 4140–4143; d) W. R. Bowman, M. R. J. Elsegood, T. Stein, G. W. Weaver, *Org. Biomol. Chem.* **2007**, *5*, 103–113; e) J. R. Duvall, F. Wu, B. B. Snider, *J. Org. Chem.* **2006**, *71*, 8579–8590; f) S. H. Shim, J. S. Kim, K. H. Son, K. H. Bae, S. S. Kang, *J. Nat. Prod.* **2006**, *69*, 400–402; g) C.-W. Jao, W.-C. Lin, Y.-T. Wu, P.-L. Wu, *J. Nat. Prod.* **2008**, 71, 1275–1279.

(122) For recent reviews on quinazoline alkaloids, see: a) J. P. Michael, *Nat. Prod. Rev.* **2004**, *21*, 650–668; b) J. P. Michael, *Nat. Prod. Rep.* **2008**, 166–187.

[123] For reviews, see: a) K. Undheim, T. Benneche, In: *Comprehensive Heterocyclic Chemistry II*, Vol. 6; Pergamon: Oxford, 1998; b) D. J. Connolly, D. Cusack, T. P. O'Sullivan, P. J. Guiry, *Tetrahedron* **2005**, *61*, 10153–10202.

[124] For some recently published syntheses of 3*H*-quinazolin-4-ones, see: a) W. Zeghida, J. Debray, S. Chierici, P. Dumy, M. Demeunynck, *J. Org. Chem.* **2008**, *73*, 2473–2475; b) S. B. Mhaske, N. P. Argade, *J. Org. Chem.* **2004**, *69*, 4563–4566.

[125] a) A. Krasovskiy, P. Knochel, *Angew. Chem.* **2004**, *116*, 3396–3399; *Angew. Chem. Int. Ed.* **2004**, *43*, 3333–3336; b) H. Ren, A. Krasovskiy, P. Knochel, *Org. Lett.* **2004**, 6, 4215–4217.

[126] Some other electrophiles, such as aldehydes, ketones and carbon dioxide have also been tested. The results are summarized in the next chapter.

[127] U. Schöllkopf, K.-W. Henneke, K. Madawinata, R. Harms, *Justus Liebigs Ann. Chem.* **1977**, 1, 40–50.

[128] a) R. Murdoch, W. R. Tully, R. Westwood, *J. Heterocycl. Chem.* **1986**, *23*, 833–841; b) B. L. Chenard, W. M. Welch, J. F. Blake, T. W. Butler, A. Reinhold, F. E. Ewing, F. S. Menniti, M. J. Pagnozzi, *J. Med. Chem.* **2001**, *44*, 1710–1717; c) Y. V. Bilokin, S. M. Kovalenko, *Heterocycl. Commun.* **2000**, 6, 409–414.

[129] a) W. R. Bowman, M. R. J. Elsegood, T. Stein, G. W. Weaver, *Org. Biomol. Chem.* **2007**, *5*, 103–113; b) R. Tangirala, S. Antony, K. Agama, D. P. Curran, *Synlett* **2005**, *18*, 2843–2846; c) C. Kaneko, K. Kasai, N. Katagiri, T. Chiba, *Chem. Pharm. Bull.* **1986**, *34*, 3672–3681.

[130] A. Perdicaro, G. Granata, A. Marrazzo, A. Santagati, *Synth. Commun.* **2008**, *38*, 723–737.

[131] For selected examples of recent syntheses of deoxyvasicinone, see: a) S. B. Mhaske, N. P. Argade, *J. Org. Chem.* **2001**, *66*, 9038–9040; b) J.-F. Liu, P. Ye, K. Sprague, K. Sargent, D. Yohannes, C. M. Baldino, C. J. Wilson, S.-C. Ng, *Org. Lett.* **2005**, *7*, 3363–3366; c) E. S. Lee, J.-G. Park, Y. Jahng, *Tetrahedron Lett.* **2003**, *44*, 1883–1886; d) A. Hamid, A. Elomri, A. Daich, *Tetrahedron Lett.* **2006**, *47*, 1777–1781; e) W. R. Bowman, M. R. J. Elsegood, T. Stein, G. W. Weaver, *Org. Biomol. Chem.* **2007**, *5*, 103–113.

[132] For recent syntheses of tryptanthrine, see: a) K. C. Jahng, S. I. Kim, D. H. Kim, C. S. Seo, J.-K. Son, S. H. Lee, E. S. Lee, Y. Jahng, *Chem. Pharm. Bull.* **2008**, *56*, 607–609; b) B. Batanero, F. Barba, *Tetrahedron Lett.* **2006**, *47*. 8201–8203.

[133] A. V. Lygin, A. de Meijere *J. Org. Chem.* **2009**, *74*, 4554–4559.

[134] a) U. Schöllkopf, F. Gerhart, I. Hoppe, R. Harms, K. Hantke, K.-H. Scheunemann, E. Eilers, E. Blume, *Justus Liebigs Ann. Chem.* **1976**, 183–202; b) W. A. Böll, A. Gerhart, A. Nürrenbach, U. Schöllkopf, *Angew. Chem.* **1970**, *82*, 482–483; *Angew. Chem. Int. Ed. Engl.* **1970**, *9*, 458–459; c) U. Schöllkopf, P. Böhmes, *Angew. Chem.* **1971**, *83*, 490–491; *Angew. Chem. Int. Ed. Engl.* **1971**, *10*, 491–492.

[135] For reviews, see: a) M. E. Jung, G. Piizzi, *Chem. Rev.* **2005**, *105*, 1735–1766; b) P. G. Sammes, D. J. Weller, *Synthesis*, **1995**, 1205–1222.

[136] For a review, see: E. V. Gromachevskaya, F. V. Kvitkovskii, T. P. Kosulina, V. G. Kul'nevich, *Chem. Heterocycl. Compd. (N.Y)* **2003**, *39*, 137–155.

[137] For a few examples of such compounds published to date, see: a) I. Fleming, M. A. Loreto, I. H. M. Wallace, *J. Chem. Soc., Perkin Trans. 1* **1986**, 349–359; b) R. R. Gataullin, I. S. Afonkin, A. A. Fatykhov, L. V. Spirikhin, E. V. Tal'vinskii, I. B. Abdrakhmanov, *Russ. Chem. Bull.* **2001**, *50*, 659–664.

[138] K. Kobayashi, S. Nagato, M. Kawakita, O. Morikawa, H. Konishi, *Chem. Lett.* **1995**, *24*, 575–576.

[139] For examples of naturally occurring 4*H*-3,1-benzoxazin-4-ones, see: a) J. J. Mason, J. Bergman, T. Janosik, *J. Nat. Prod.* **2008**, *71*, 1447–1450; b) H. Wang, A. Ganesan, *J. Org. Chem.* **1998**, *63*, 2432–2433.

[140] a) G. Fenton, C. G. Newton, B. M. Wyman, P. Bagge, D. I. Dron, D. Riddell, G. D. Jones, *J. Med. Chem.*, **1989**, *32*, 265–272; b) L. Hedstrom, A. R. Moorman, J. Dobbs, R. H. Abeles, *Biochemistry*, **1984**, *23*, 1753–1759; For reviews on 4*H*-3,1-benzoxazin-4-ones, see: c) G. M. Coppola, *J. Heterocycl. Chem.* **1999**, *36*, 563–588; d) G. M. Coppola, *J. Heterocycl. Chem.* **2000**, *37*, 1369–1388.

[141] a) K. Ding, Y. Lu, Z. Nikolovska-Coleska, G. Wang, S. Qiu, S. Shangary, W. Gao, D. Qin, J. Stuckey, K. Krajewski, P. P. Roller, S. Wang, *J. Med. Chem.* **2006**, *49*, 3432–3435; b) T. Jiang, K. L. Kuhen, K. Wolff, H. Yin, K. Bieza, J. Caldwell, B. Bursulaya, T. Y.-H. Wu, Y. He, *Bioorg. Med. Chem. Lett.* **2006**, *16*, 2105–2108; c) K. C. Luk, S. S. So, J. Zhang, Z. Zhang, (F. Hoffman-LaRoche AG), WO 2006/136606 A3, 2006; d) P. Hewawasam, V. K. Gribkoff, Y. Pendri, S. I. Dworetzky, N. A. Meanwell, E. Martinez, C. G. Boissard, D. J. Post-Munson, J. T. Trojnacki, K. Yeleswaram, L. M. Pajor, J. Knipe, Q. Gao, R. Perrone, J. E., Jr. Starrett, *Bioorg. Med. Chem. Lett.* **2002**, *12*, 1023–1026; e) R. Sarges, H. R. Howard, K. Koe, A. Weissman, *J. Med. Chem.* **1989**, *32*, 437–444.

[142] For some syntheses of iminophthalanes, see: a) R. Sato, M. Ohmori, F. Kaitani, A. Kurosawa, T. Senzaki, T. Goto, M. Saito, *Bull. Chem. Soc. Jpn.* **1988**, *61*, 2481–2485; b) H. Suzuki, M. Koge, A. Inoue, T. Hanafusa, *Bull. Chem. Soc. Jpn.* **1978**, *51*, 1168–1171; c) H. Suzuki, M. Koge, T. Hanafusa, *J. Chem. Soc. Chem. Commun.* **1977**, 341–342.

[143] a) D. Lednicer, E. D. Emmert, *J. Heterocycl. Chem.* **1970**, *7*, 575–581; b) R. R. Schmidt, B. Beitzke, *Chem. Ber.* **1983**, *116*, 2115–2135.

[144] For reviews on [1,2]-Wittig rearrangement, see: a) K. Tomooka, H. Yamamoto, T. Nakai, *Liebigs Ann./Recueil* **1997**, 1275–1281; b) K. Tomooka, In *The Chemistry of OrganolithiumCompounds*; Rappoport, Z., Marek, I., Eds.; Wiley: London, 2004; Vol. 2, pp 749–828.

[145] For reviews, see: a) J. A. Vanecko, H. Wan, F. G. West, *Tetrahedron* **2006**, *62*, 1043–1062; b) I. E. Markó, B. M. Trost, I. Fleming, Eds.; In *Comprehensive Organic Synthesis*; Pergamon: Oxford, 1991; Vol. 3, pp 913–973.

[146] The author is grateful to one of the referees who pointed out this possibility during the submition of the article (*J. Org. Chem.*).

[147] a) T. Saegusa, Y. Ito, *Synthesis*, **1975**, 291–300; b) T. Saegusa, Y. Ito, N. Takeda, K. Hirota, *Tetrahedron Lett.* **1967**, *8*, 1273–1275; c) T. Saegusa, Y. Ito, S. Kobayashi, K. Hirota, *Tetrahedron Lett.* **1967**, *8*, 521–524.

[148] A mixture of **91** in benzene with 10 mol% of Cu_2O was heated under reflux for 1 h. No changes were detected according to TLC.

[149] A. V. Lygin, A. de Meijere, *Eur. J. Org. Chem.* **2009**, 5138–5141.

[150] For reviews, see: a) A. Kamal, K. L. Reddy, V. Devaiah, N. Shankaraiah, M. V. Rao, *Mini Rev. Med. Chem.* **2006**, *6*, 71–89; b) R. R. Wexler, W. J. Greenlee, J. D. Irvin, M. R.

Goldberg, K. Prendergast, R. P. Smith, P. B. M. W. M. Timmermans, *J. Med. Chem.* **1996**, *39*, 625–656.

[151] a) Y.-F. Li, G.-F. Wang, P.-L. He, W.-G. Huang, F.-H. Zhu, H.-Y. Gao, W. Tang, Y. Luo, C.-L. Feng, L.-P. Shi, Y.-D. Ren, W. Lu, J.-P. Zuo, *J. Med. Chem.* **2006**, *49*, 4790–4794; b) P. L. Beaulieu, Y. Bousquet, J. Gauthier, J. Gillard, M. Marquis, G. McKercher, C. Pellerin, S. Valois, G. Kukolj, *J. Med. Chem.*, **2004**, *47*, 6884–6892; c) S. Hirashima, T. Suzuki, T. Ishida, S. Noji, S. Yata, I. Ando, M. Komatsu, S. Ikeda, H. Hashimoto, *J. Med. Chem.* **2006**, *49*, 4721–4736.

[152] M. Sabat, J. C. Vanrens, M. J. Laufersweiler, T. A. Brugel, J. Maier, A. Golebiowski, B. De, V. Easwaran, L. C. Hsieh, R. L. Walter, M. J. Mekel, A. Evdokimov, M. J. Janusz, *Bioorg. Med. Chem. Lett.* **2006**, *16*, 5973–5977.

[153] N. H. Hauel, H. Nar, H. Priepke, U. Ries, J. Stassen, W. Wienen, *J. Med. Chem.* **2002**, *45*, 1757–1766.

[154] H. Nakano, T. Inoue, N. Kawasaki, H. Miyataka, H. Matsumoto, T. Taguchi, N. Inagaki, H. Nagai, T. Satoh, *Bioorg. Med. Chem.* **2000**, *8*, 373–380.

[155] a) P. N. Preston, *Benzimidazoles and Congeneric Tricyclic Compounds*. In *The Chemistry of Heterocyclic Compounds* (Eds.: A. Weissberger, E. C. Taylor), Wiley-VCH, New York, **1981**, vol. 40, pp. 6–60; b) M. R. Grimmett, *Imidazoles and their Benzo Derivatives*. In *Comprehensive Heterocyclic Chemistry* (Eds.: A. R. Katritzky, C. W. Rees), Pergamon, Oxford, **1984**, vol. 5, pp. 457–487.

[156] For catalytic syntheses of benzimidazoles and related cyclizations, see: a) S. Murru, B. K. Patel, J. Le Bras, J. Muzart, *J. Org. Chem.* **2009**, *74*, 2217–2220; b) Z. Li, H. Sun, H. Jiang, H. Liu, *Org. Lett.* **2008**, *10*, 3263–3266; c) G. Brasche, S. L. Buchwald, *Angew. Chem.* **2008**, *120*, 1958–1960; *Angew. Chem. Int. Ed.* **2008**, *47*, 1932–1934; d) J. Alen, K. Robeyns, W. M. De Borggraeve, L. Van Meervelt, F. Compernolle, *Tetrahedron*, **2008**, *64*, 8128–8133; e) B. Zou, Q. Yuan, D. Ma, *Angew. Chem.* **2007**, *119*, 2652–2655; *Angew. Chem. Int. Ed.* **2007**, *46*, 2598–2601; f) C. Venkatesh, G. S. M. Sundram, H. Ila, H. Junjappa, *J. Org. Chem.* **2006**, *71*, 1280–1283; g) K. G. Nazarenko, T. I. Shyrokaya, K. V. Shvidenko, A. A. Tolmachev, *Synth. Commun.* **2003**, *33*, 4303–4311; h) G. Evindar, R. A. Batey, *Org. Lett.* **2003**, *5*, 133–136; i) C. T. Brain, J. T. Steer, *J. Org. Chem.* **2003**, *68*, 6814–6816; g) C. T. Brain, S. A. Brunton, *Tetrahedron Lett.* **2002**, *43*, 1893–1895.

[157] T. Saegusa, Y. Ito, K. Kobayashi, K. Hirota, H. Yoshioka, *Tetrahedron. Lett.* **1966**, 6121–6124.

[158] T. Saegusa, Y. Ito, K. Kobayashi, K. Hirota, H. Yoshioka, *Bull. Chem. Soc. Jpn.* **1969**, *42*, 3310–3313.

[159] For recent reviews on copper-catalyzed cross-coupling reactions, see: a) S. V. Ley, A. W. Thomas, *Angew. Chem.,* **2003**, *115*, 5558–5607; *Angew. Chem., Int. Ed.* **2003**, *42*, 5400–5449; b) K. Kunz, U. Scholz, D. Ganzer, *Synlett* **2003**, 2428–2439; c) I. P. Beletskaya, A. V. Cheprakov, *Coord. Chem. Rev.* **2004**, *248*, 2337–2364.

[160] a) For a review, see: H. W. Gschwend, H. R. Rodriquez, *Org. React. (N. Y.)*, **1979**, *26*, 1–360; b) A. R. Katritzky, W. H. Ramer, J. N. Lam, *J. Chem. Soc, Perkin Trans. 1* **1987**, 775–780.

[161] For relevant reviews on transition-metal-catalyzed C-C bond formation via C-H bond cleavage, see: a) D. A. Colby, R. G. Bergman, J. A. Ellman, *Chem. Rev.* **2009**, Article ASAP; b) J. C. Lewis, R. G. Bergman, J. A. Ellman, *Acc. Chem. Res.* **2008**, *41*, 1013–1025; c) F. Kakiuchi, T. Kochi, *Synthesis* **2008**, 3013–3039; d) D. Alberico, M. E. Scott, M. Lautens, *Chem. Rev.* **2007**, *107*, 174–238; e) I. V. Seregin, V. Gevorgyan, *Chem. Soc. Rev.* **2007**, *36*, 1173–1193.

[162] a) J. Ezquerra, C. Lamas, *Tetrahedron* **1997**, *53*, 12755–12764; b) Y. Gong, W. He, *Org. Lett.* **2002**, *4*, 3803–3805; c) B. B. Wang, P. J. Smith, *Tetrahedron Lett.* **2003**, *44*, 8967–8969; d) M. W. Hooper, M. Utsunomiya, J. F. Hartwig, *J. Org. Chem.* **2003**, *68*, 2861–2873.

[163] a) A. G. Mistry, K. Smith, M. R. Bye, *Tetrahedron Lett.* **1986**, *27*, 1051–1054; b) I. Kawasaki, N. Taguchi, Y. Yoneda, M. Yamashita, S. Ohta, *Heterocycles* **1996**, *43*, 1375–1379.

[164] The same benzimidazole **217n** was also observed as a major product in the reaction of **147-Br** with 2,4,6-trimethylaniline.

[165] R. Obrecht, R. Herrmann, I. Ugi, *Synthesis*, **1985**, 400–402.

[166] G. Bengtson, S. Keyaniyan, A. de Meijere, *Chem. Ber.* **1986**, *119*, 3607–3630.

[167] R. S. Bon, B. van Vliet, N. E. Sprenkels, R. F. Schmitz, F. J. J. de Kanter, C. V. Stevens, M. Swart, F. M. Bickelhaupt, M. B. Groen, R. V. A. Orru, *J. Org. Chem.* **2005**, *70*, 3542–3553.

[168] K. L. Stevens, D. K. Jung, M. J. Alberti, J. G. Badiang, G. E. Peckham, J. M. Veal, M. Cheung, P. A. Harris, S. D. Chamberlain, M. R. Peel, *Org. Lett.* **2005**, *7*, 4753–4756.

[169] I. J. Solomon, R. Filler, *J. Am. Chem. Soc.* **1963**, *85*, 3492–3496.

[170] D. H. Wadsworth, S. M. Geer, M. R. Detty, *J. Org. Chem.* **1987**, 52, 3662–3668.

[171] J. Cossy, J.-P. Pete, *Bull. Soc. Chim. Fr.* **1979**, 559–567.

[172] M. E. Pierce, R. L. Parsons, L. A. Radesca, Y. S. Lo, S. Silverman, *J. Org. Chem.* **1998**, *63*, 8536–8543.

[173] C. Housseman, J. Zhu, *Synlett*, **2006**, 1777–1779.

[174] U. Schöllkopf, R. Schröder, D. Stafforst, *Liebigs Ann. Chem.* **1974**, 44–53.

[175] L. J. Goossen, M. Blanchot, C. Brinkmann, K. Goossen, R. Karch, A. Rivas-Nass, *J. Org. Chem.* **2006**, 71, 9506–9509.

[176] R. Moffett, *J. Med. Chem.* **1972**, *15*, 1079–1081.

[177] A. Krantz, B. Hoppe, *Tetrahedron Lett.* **1975**, 9; 695–698.

[178] J. M. Barker, P. R. Huddleston, M. L. Wood, *Synth. Commun.* **1995**, 3729–3734.

[179] X. E. Hu, J. M. Cassady, *Synth. Commun.* **1995**, *25*, 907–913.

[180] W. C. Still, M. Kahn, A. Mitra, *J. Org. Chem.* **1978**, *43*, 2923–2925.

[181] M. Suzuki, M. Miyoshi, K. Matsumoto, *J. Org. Chem.* **1974**, *39*, 1980–1980.

[182] H. Saikachi, T. Kitagawa, H. Sasaki, *Chem. Pharm. Bull.* **1979**, *27*, 2857–2861.

[183] A. Padwa, J. R. Gasdaska, G. Hoffmanns, H. Rebello, *J. Org. Chem.* **1987**, *52*, 1027–1035.

[184] R. J. Cherney, P. Carter, J. V. Duncia, D. S. Gardner, J. B. Santella (Bristol-Myers Squibb Company), WO 2004071460 A2 20040826, **2004**.

[185] H. Horikawa, T. Iwasaki, K. Matsumoto, M. Miyoshi, *J. Org. Chem.* **1978**, *43*, 335–337.

[186] J. H. Rigby, S. Laurent, *J. Org. Chem.* **1998**, *63*, 6742–6744.

[187] K. Kobayashi, T. Nakashima, M. Mano, O. Morikawa, H. Konishi, *Chem. Lett.* **2001**, *7*, 602–603.

[188] P. Hrvatin, A. G. Sykes, *Synlett* **1997**, *9*, 1069–1070.

[189] B. Scherrer, *J. Org. Chem.* **1972**, *37*, 1681–1685.

[190] B. V. Lingaiah, G. Ezikiel, T. Yakaiah, G. V. Reddy, P. S. Rao, *Synlett* **2006**, *15*, 2507–2509.

[191] L. Wang, J. Xia, F. Qin, C. Qian, J. Sun, *Synthesis* **2003**, *8*, 1241–1247.

[192] D. E. Ames, S. Chandrasekhar, R. Simpson, *J. Chem. Soc., Perkin Trans. 1*, **1975**; 2035–2036.

[193] R. C. R. Gilbert, *Seances Acad. Sci. C* **1974**, *279*, 159.

[194] M. Pesson, D. C. R. Richer, *Hebd. Seances Acad. Sci.* **1965**, *260*, 603–605.

[195] J. F. Liu, P. Ye, K. Sprague, K. Sargent, D. Yohannes, C. M. Baldino, C. J. Wilson, S. C. Ng, *Org. Lett.* **2005**, *7*, 3363–3366.

[196] J. Bergman, J.-O. Lindstroem, U. Tilstam, *Tetrahedron* **1985**, *41*, 2879–2882.

[197] J. C. Sheehan, J. W. Frankenfeld, *J. Am. Chem. Soc.* **1961**, *83*, 4792–4795.

[198] M. Gütschow, *J. Org. Chem.* **1999**, *64*, 5109–5115.

[199] E. C. Wagner, U. F. Fegley, *Org. Synth.* **1955** *Coll. Vol. 3*, 488; *Org. Synth.* **1947**, *Ann. Vol. 27*, 45.

[200] J. C. Barrish, S. D. Kimball, J. Krapcho, US 4946820, 7.08.1990, Application: US 89-334025.

[201] Z.-G. Le, Z.-C. Chen, Y. Hu, Q.-G. Zheng, *Synthesis* **2004**, 208–212.

[202] J. T. Ralph, *Synth. Commun.* **1989**, *19*, 1381–1387.

[203] J. Torres, J. L. Lavandera, P. Cabildo, R. M. Claramunt, J. Elguero, *J. Heterocycl. Chem.* **1988**, *25*, 771–782.

[204] S. Benard, L. Neuville, J. Zhu, *J. Org. Chem.* **2008**, *73*, 6441–6444.

[205] E. Cuevas-Yanez, J. M. Serrano, G. Huerta, J. M. Muchowski, R. Cruz-Almanza, *Tetrahedron* **2004**, *60*, 9391–9396.

[206] B. D. Palmer, J. B. Smaill, M. Boyd, D. H. Boschelli, A. M. Doherty, J. M. Hamby, S. S. Khatana, J. B. Kramer, A. J. Kraker, R. L. Panek, G. H. Lu, T. K. Dahring, R. T. Winters, H. D. H. Showalter, W. A. Denny, *J. Med. Chem.* **1998**, *41*, 5457–5465.

F. Representative ^{1}H and ^{13}C Spectra of the prepared compounds

Dimethyl 3-(4-Ethoxyphenyl)-1*H*-pyrrole-2,4-dicarboxylate (173ac)

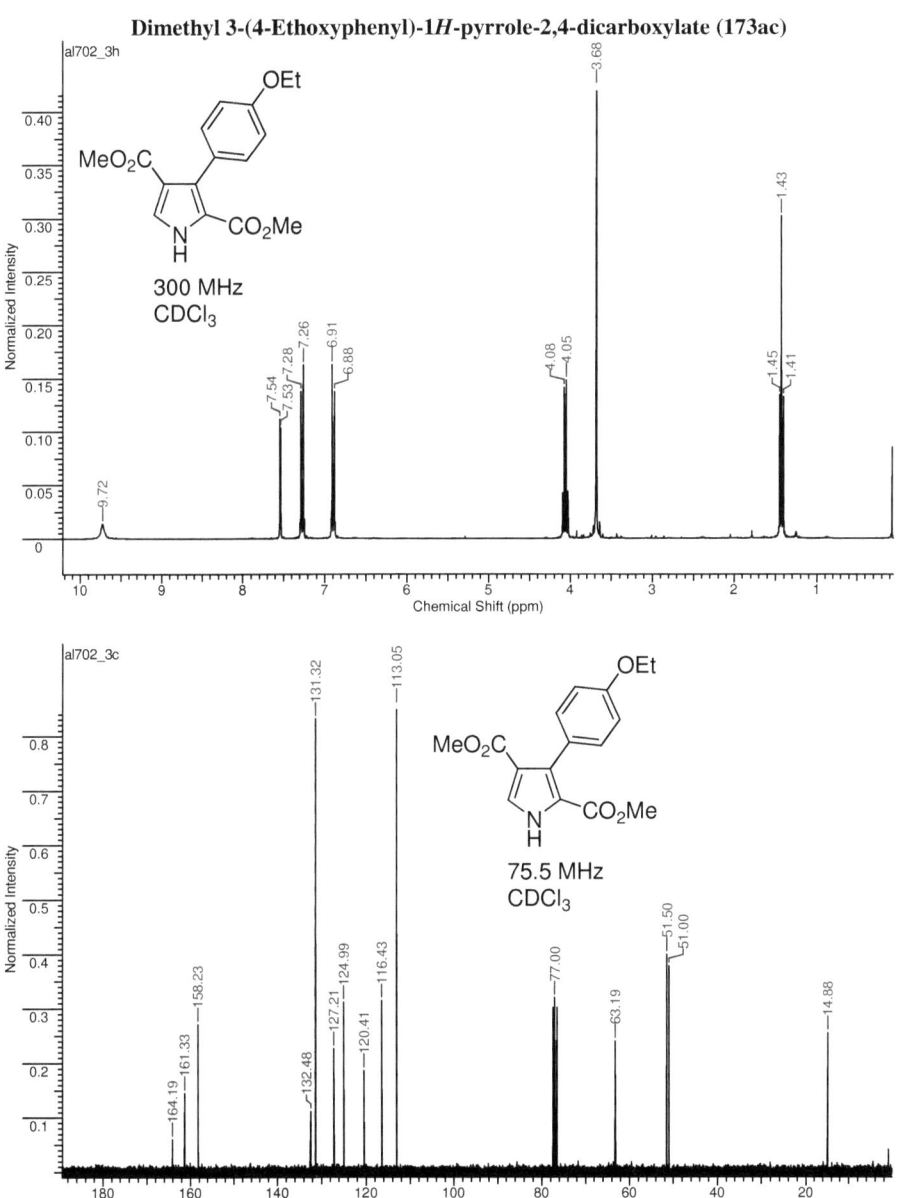

Dimethyl 3-(Thiophen-2-yl)-1H-pyrrole-2,4-dicarboxylate (173ag)

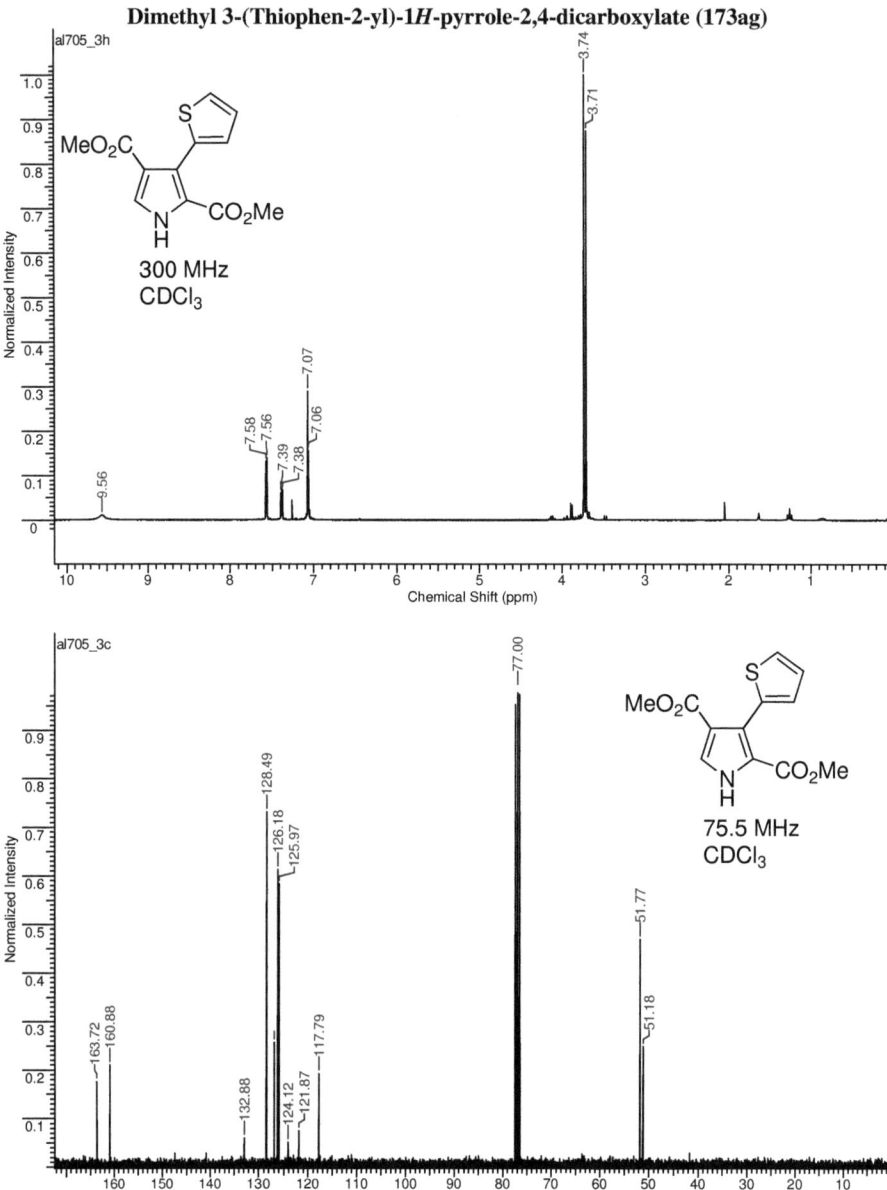

Methyl 2-Phenyl-1*H*-pyrrole-4-carboxylate (173eh)

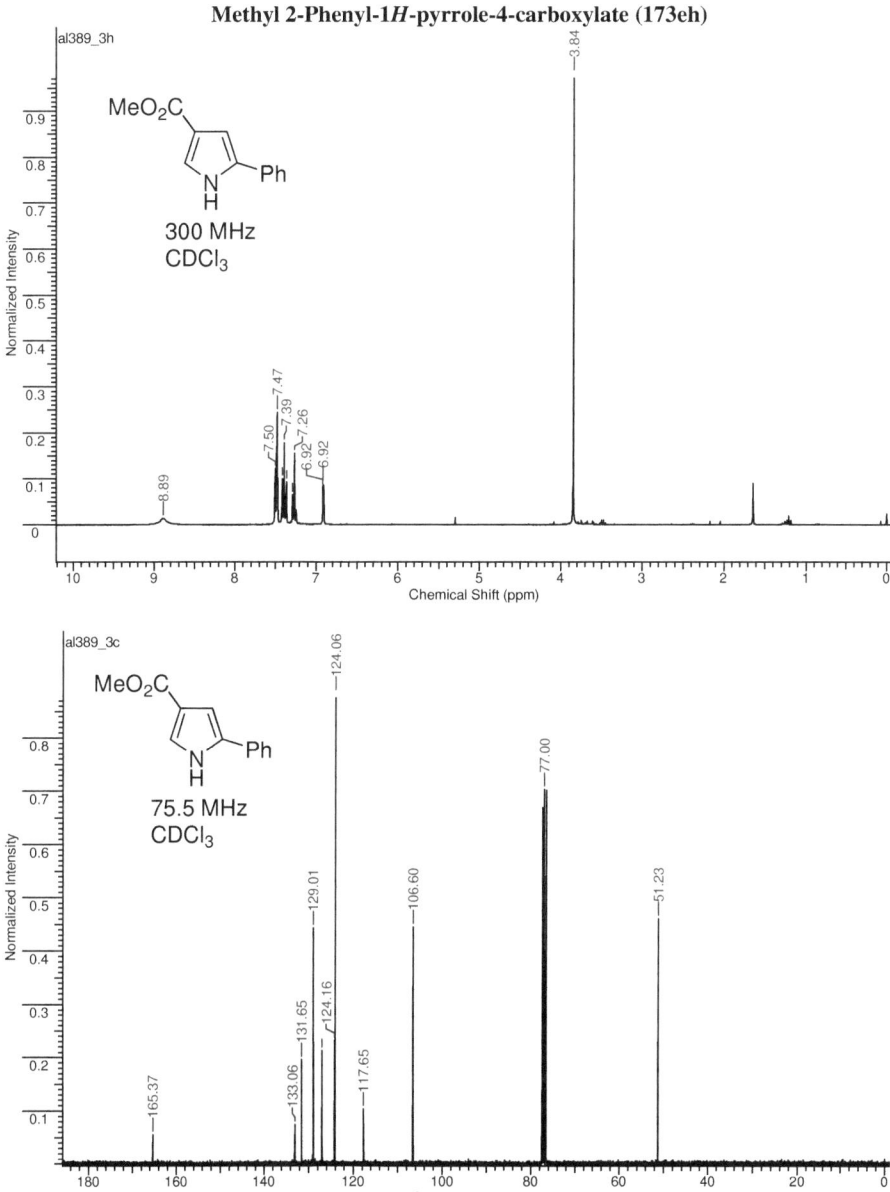

Methyl 2-(Ethoxycarbonyl)-1H-pyrrole-4-carboxylate (173bh)

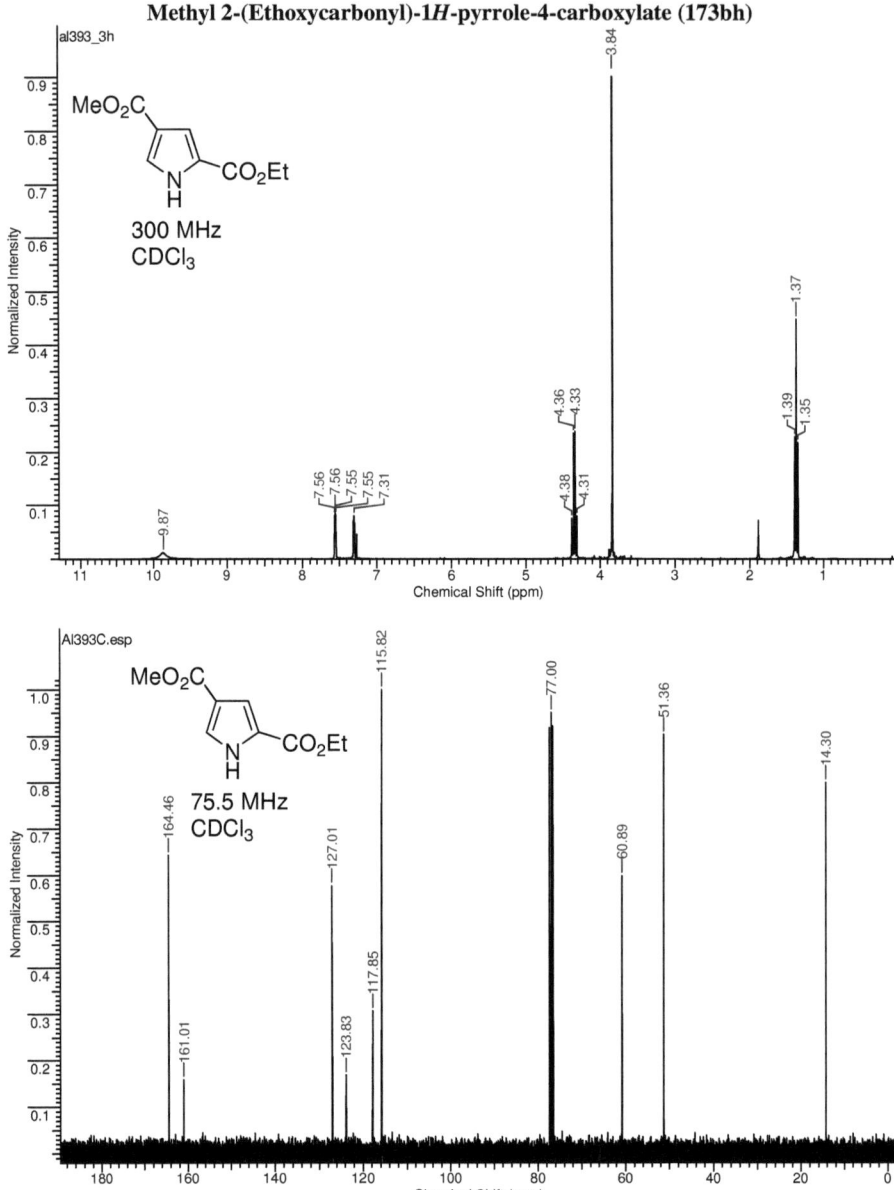

Ethyl 3-(*sec*-Butyl)-1*H*-pyrrole-2-carboxylate (178bh)

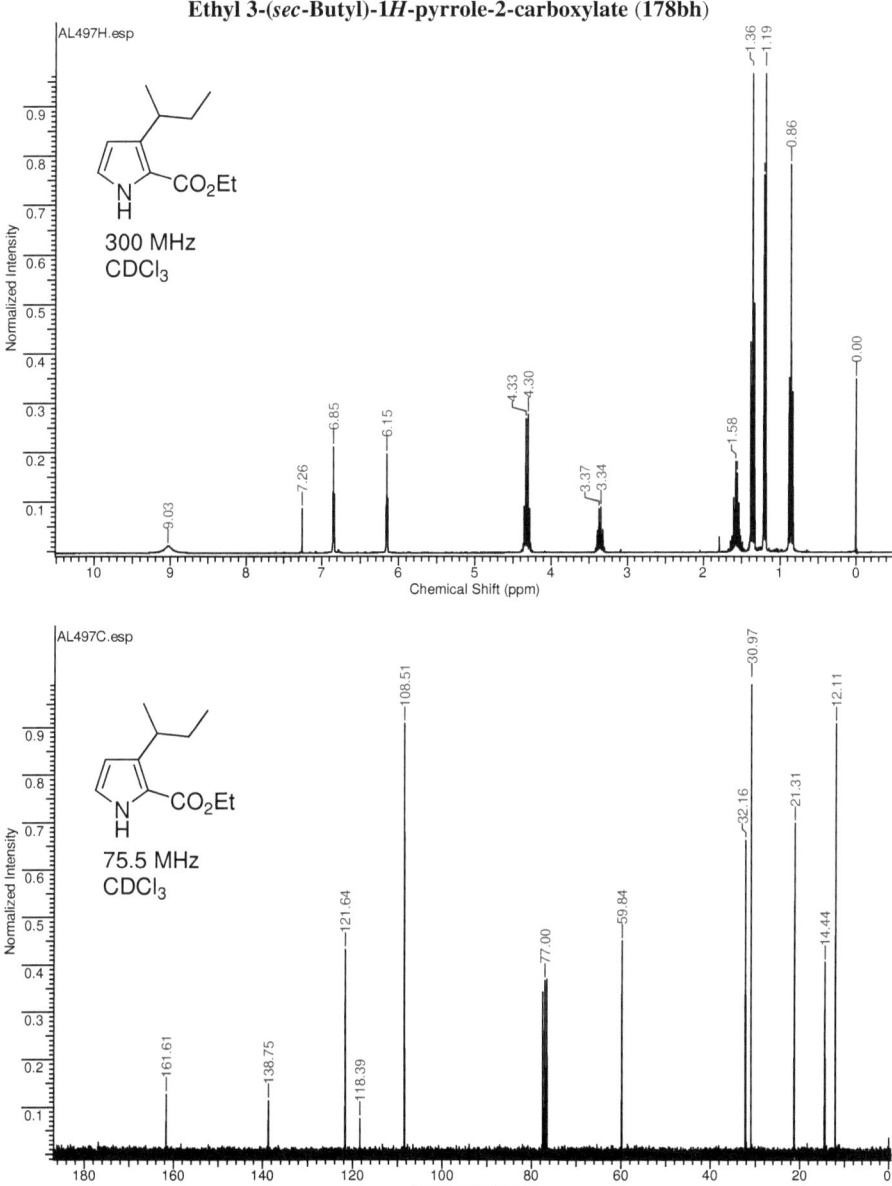

Ethyl 3-Cyclopropyl-1*H*-pyrrole-2-carboxylate (178be)

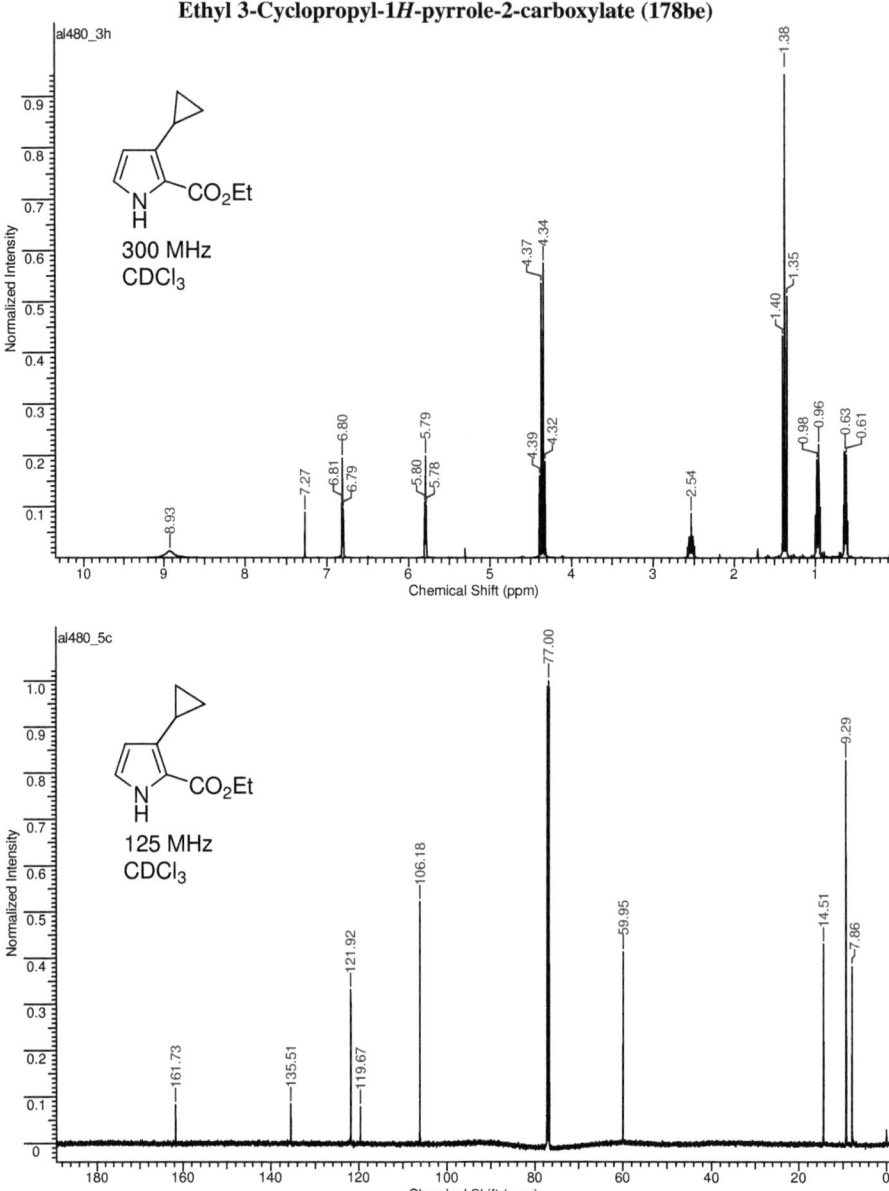

4,5-Dihydro-1-*H*-pyrano[3,4-*b*]pyrrol-7-one (179)

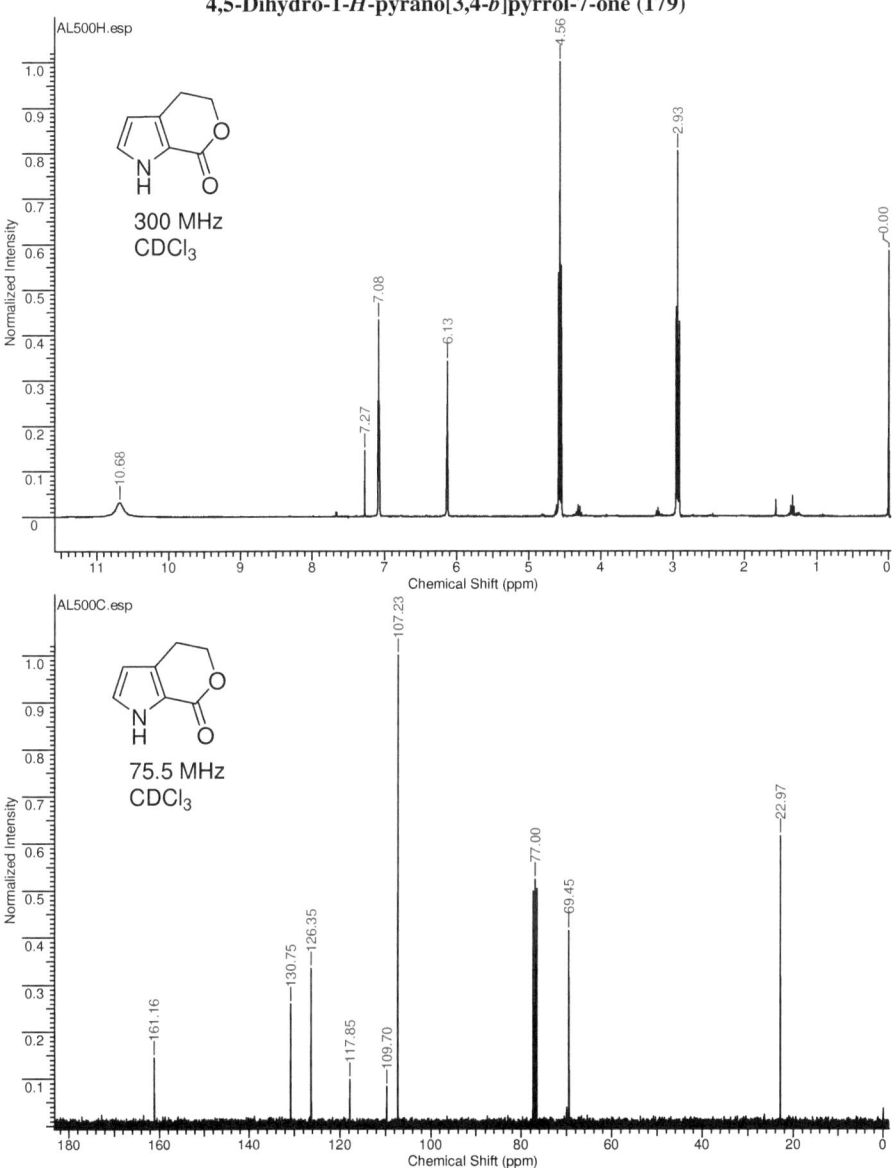

2-Isocyanobenzaldehyde (202c)

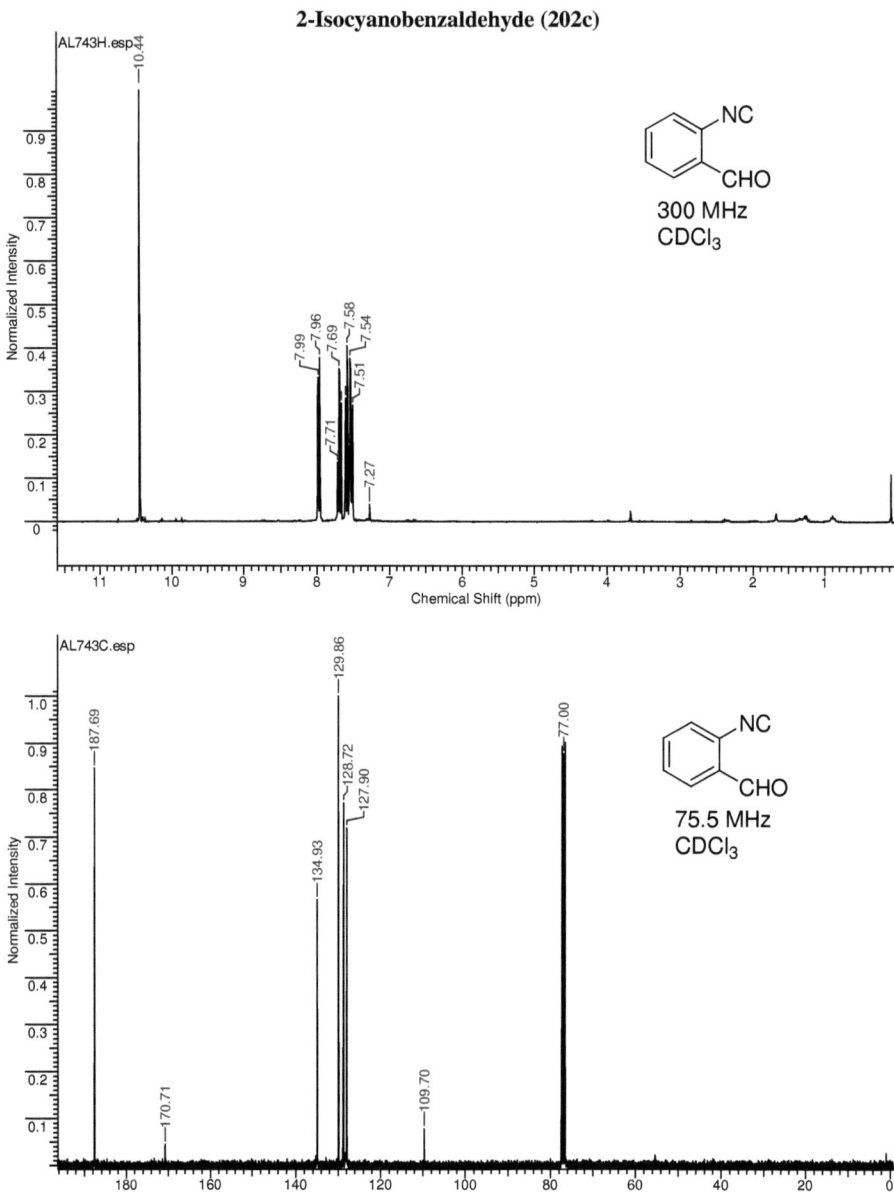

Methyl 2-isocyanobenzoate (202a)

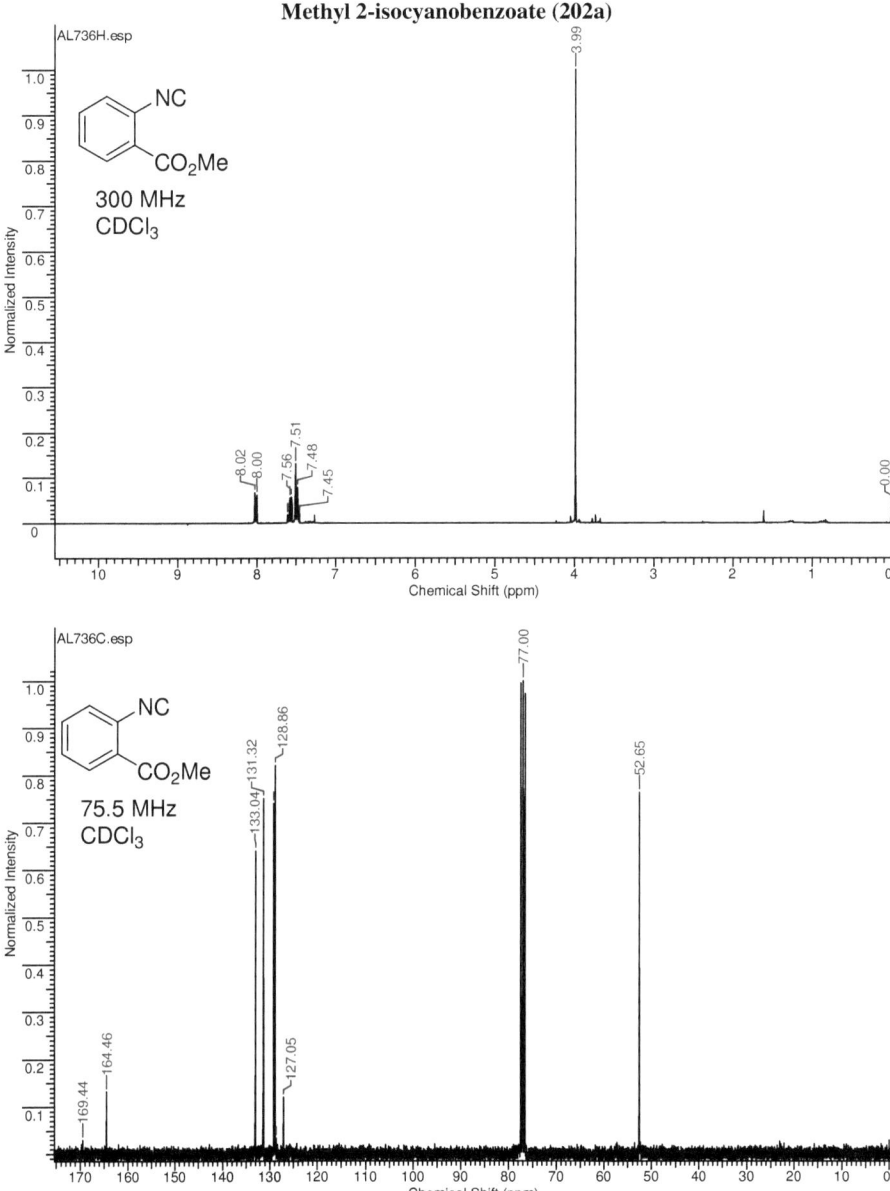

(2-Isocyanophenyl)(pyridin-4-yl)methanol (204d)

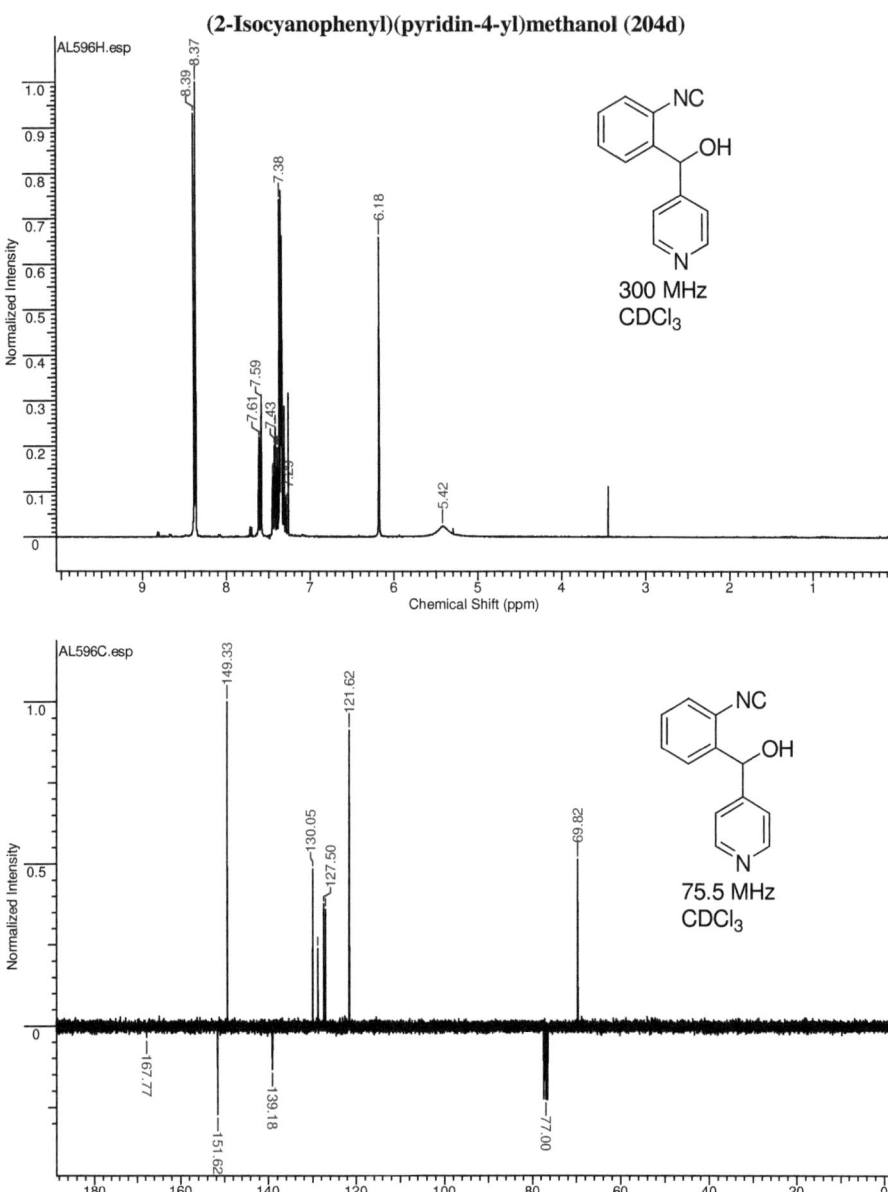

3-*p*-Tolylquinazolin-4(3*H*)-one (191b)

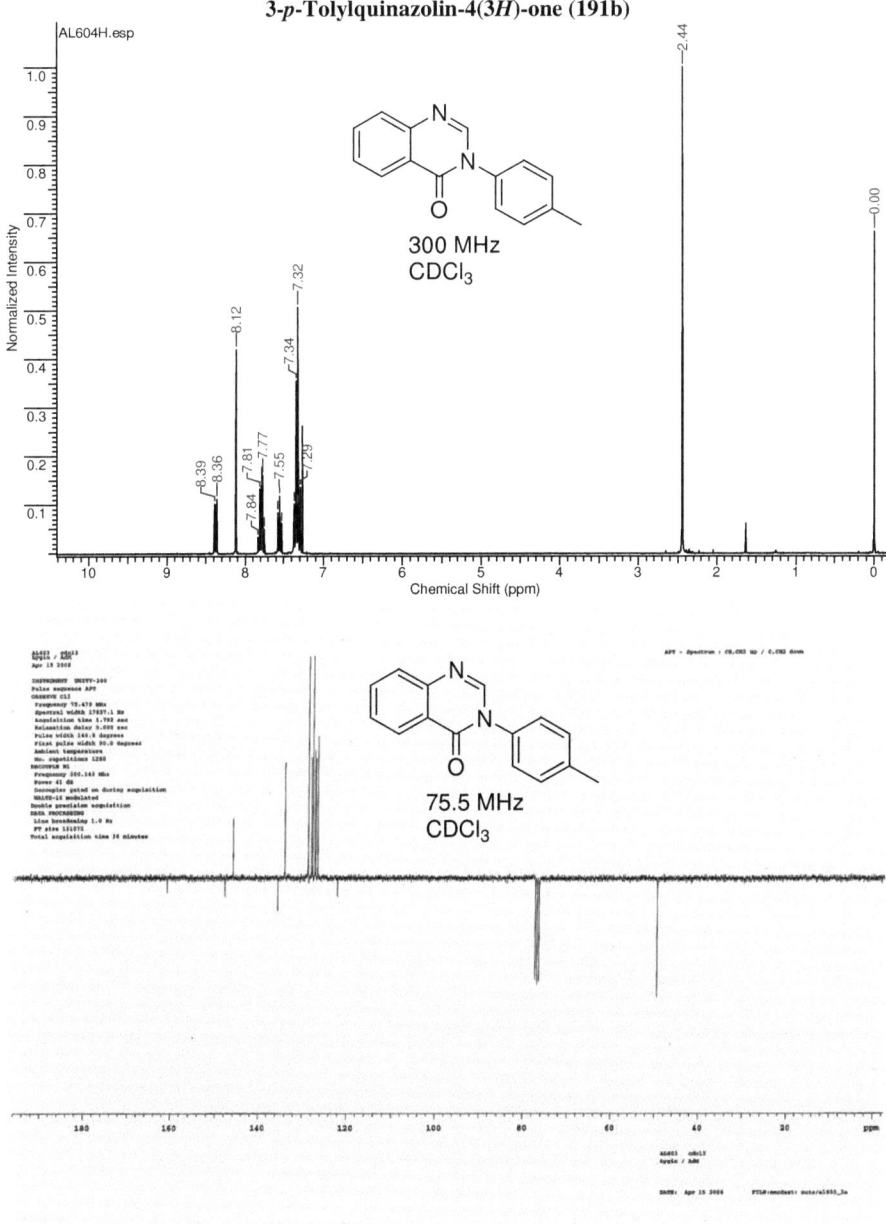

3-Isopropylquinazolin-4(3*H*)-one (191f)

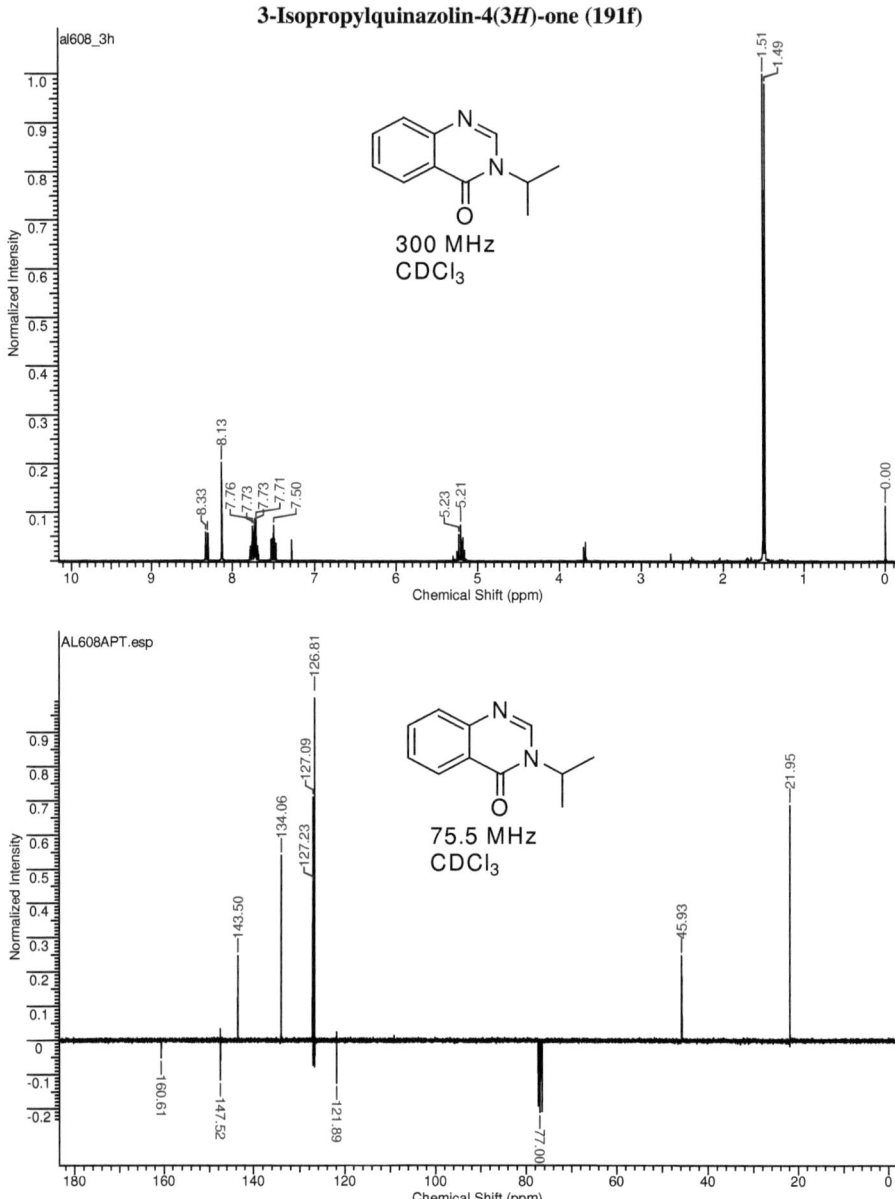

3-Cyclopropylquinazolin-4(3H)-one (191g)

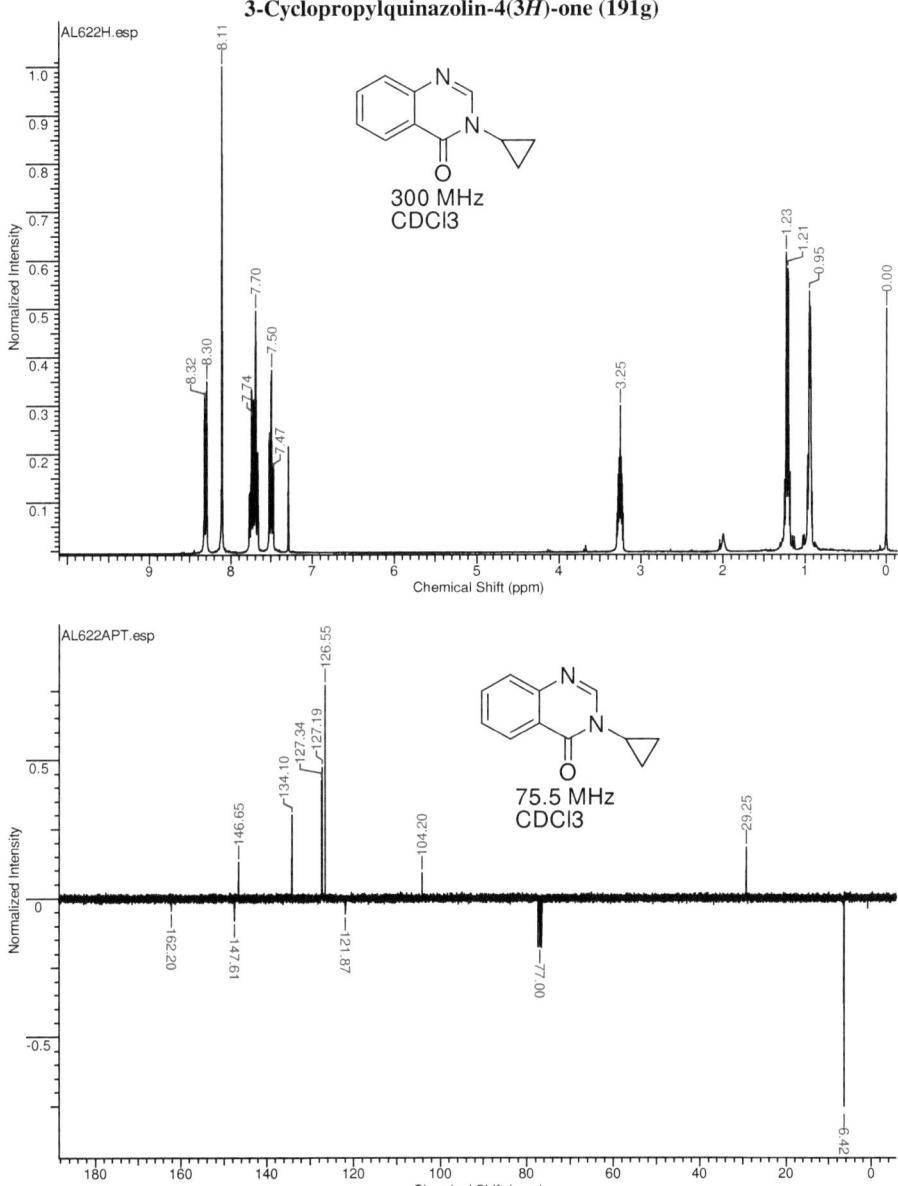

3-Cyclohexylquinazoline-4(3H)-thione (191i)

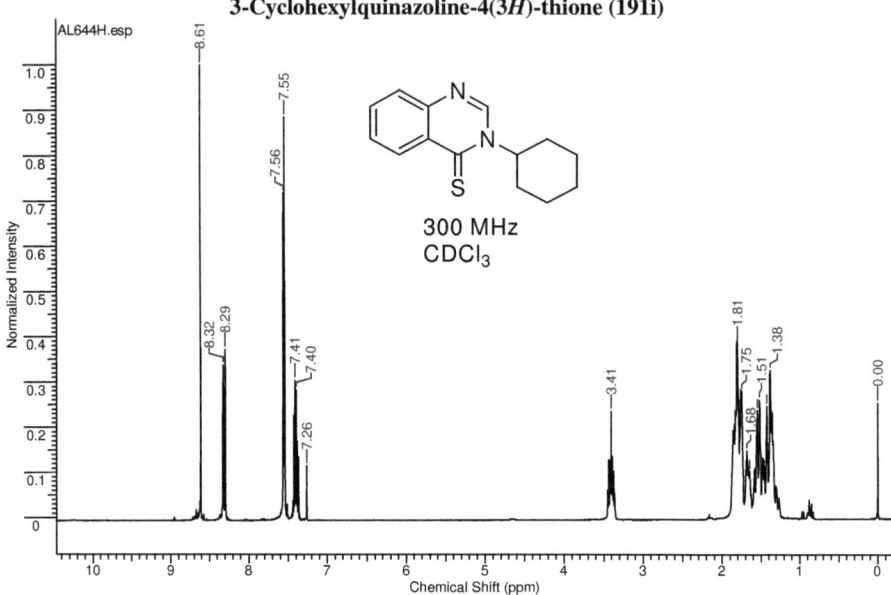

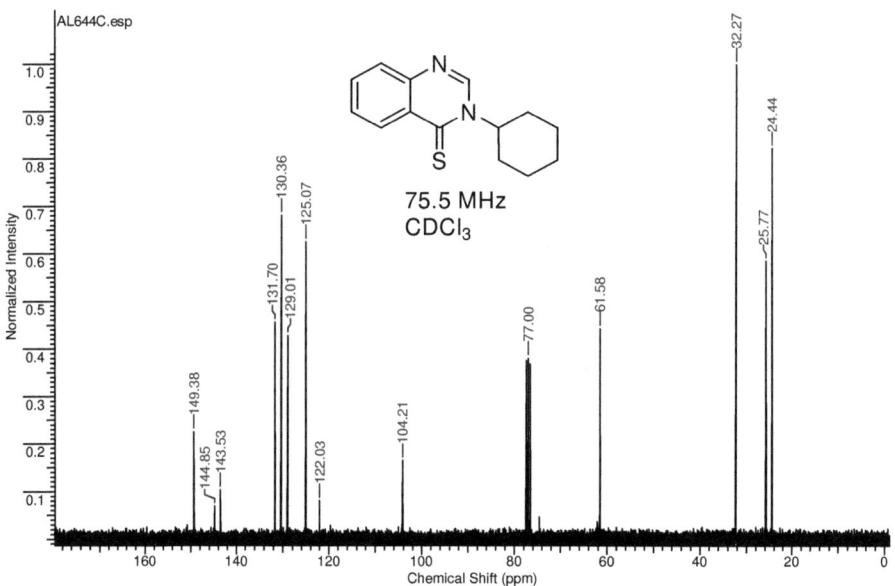

2,3-Dihydropyrrolo[2,1-b]quinazolin-9(1H)-one (desoxyvascinone, 191n)

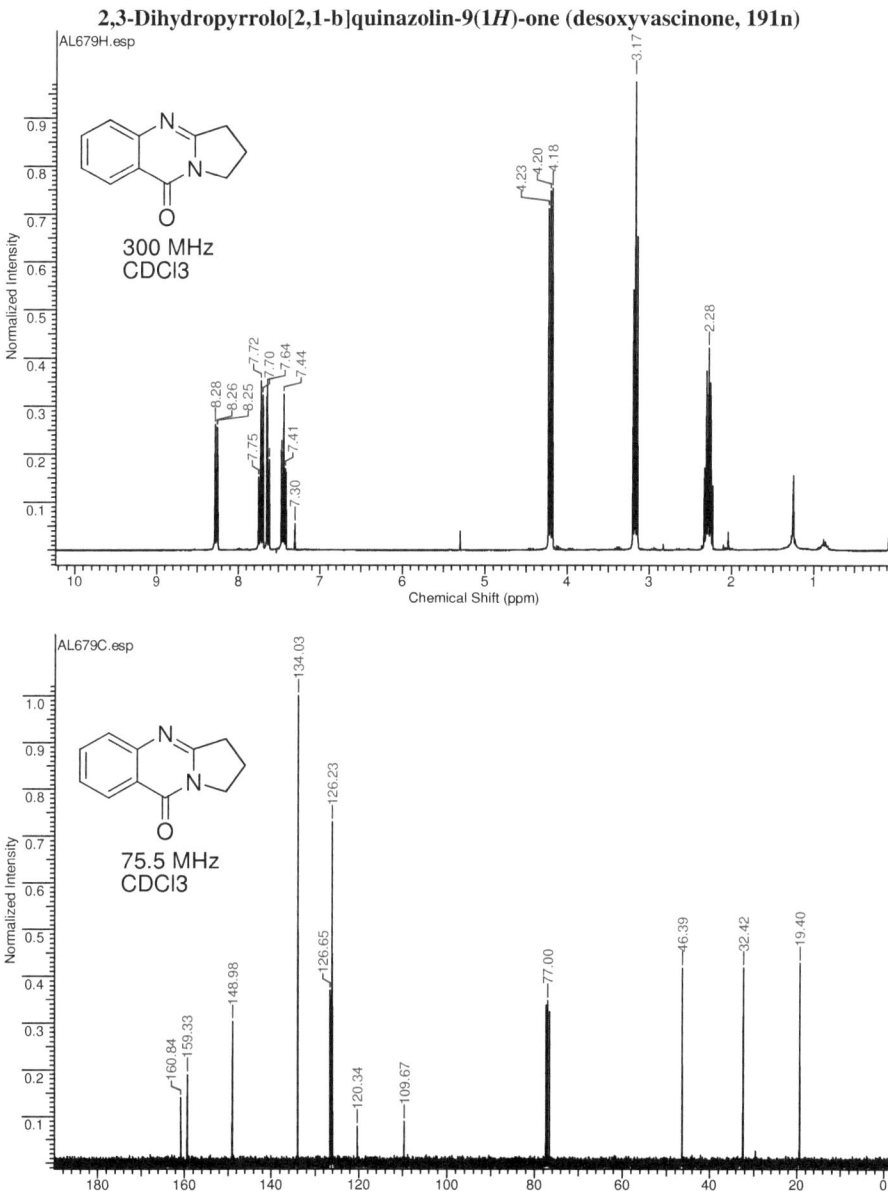

Indolo[2,1-b]quinazoline-6,12-dione (trypthamine, 191o)

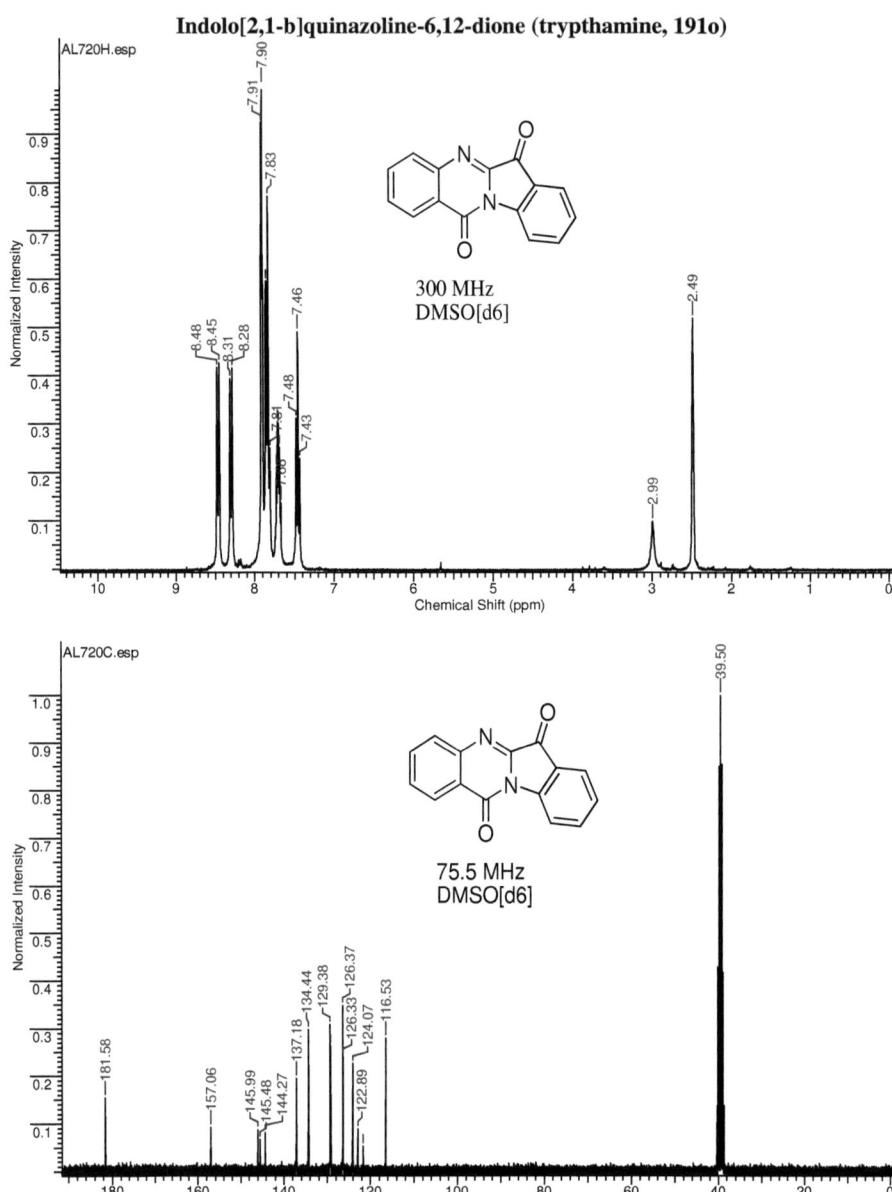

3-Benzyl-2-iodoquinazolin-4(3H)-one (191m)

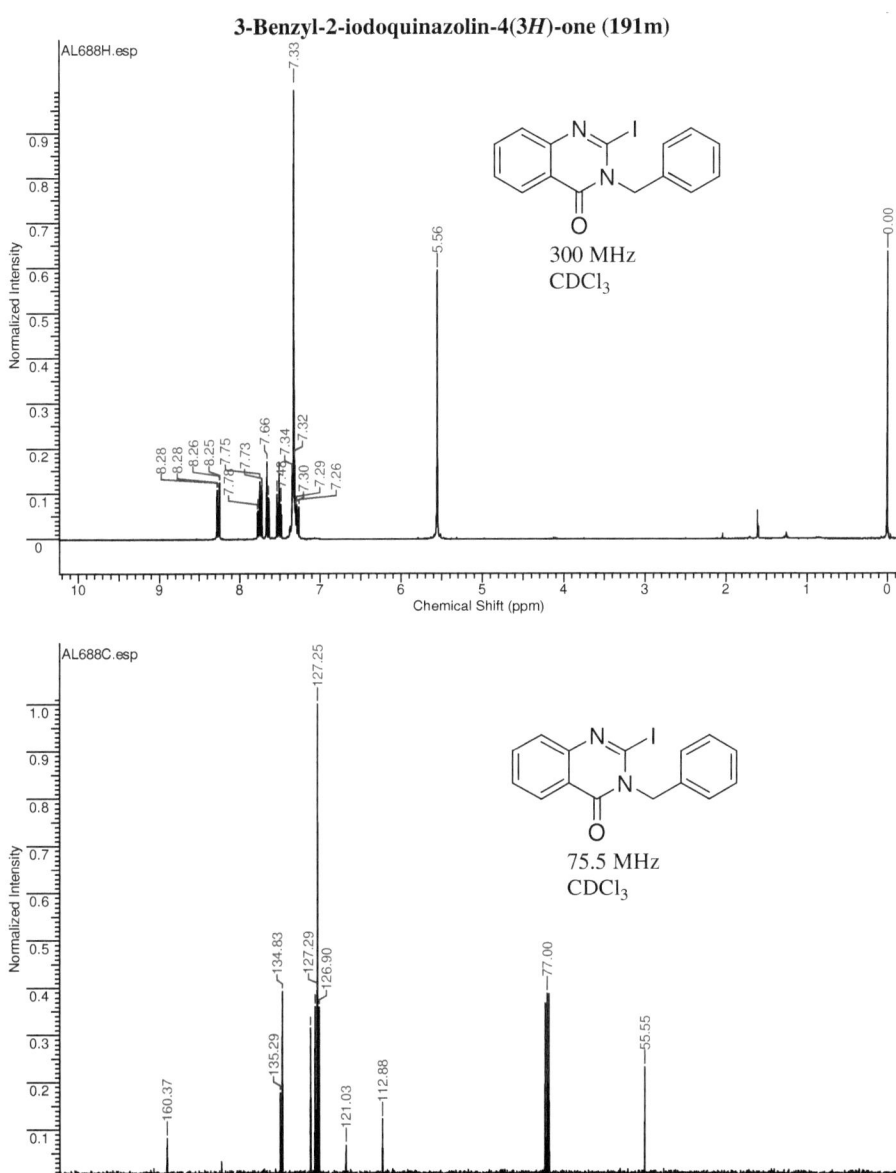

4-(Trifluoromethyl)-4-phenyl-4H-3,1-benzoxazine (201l):

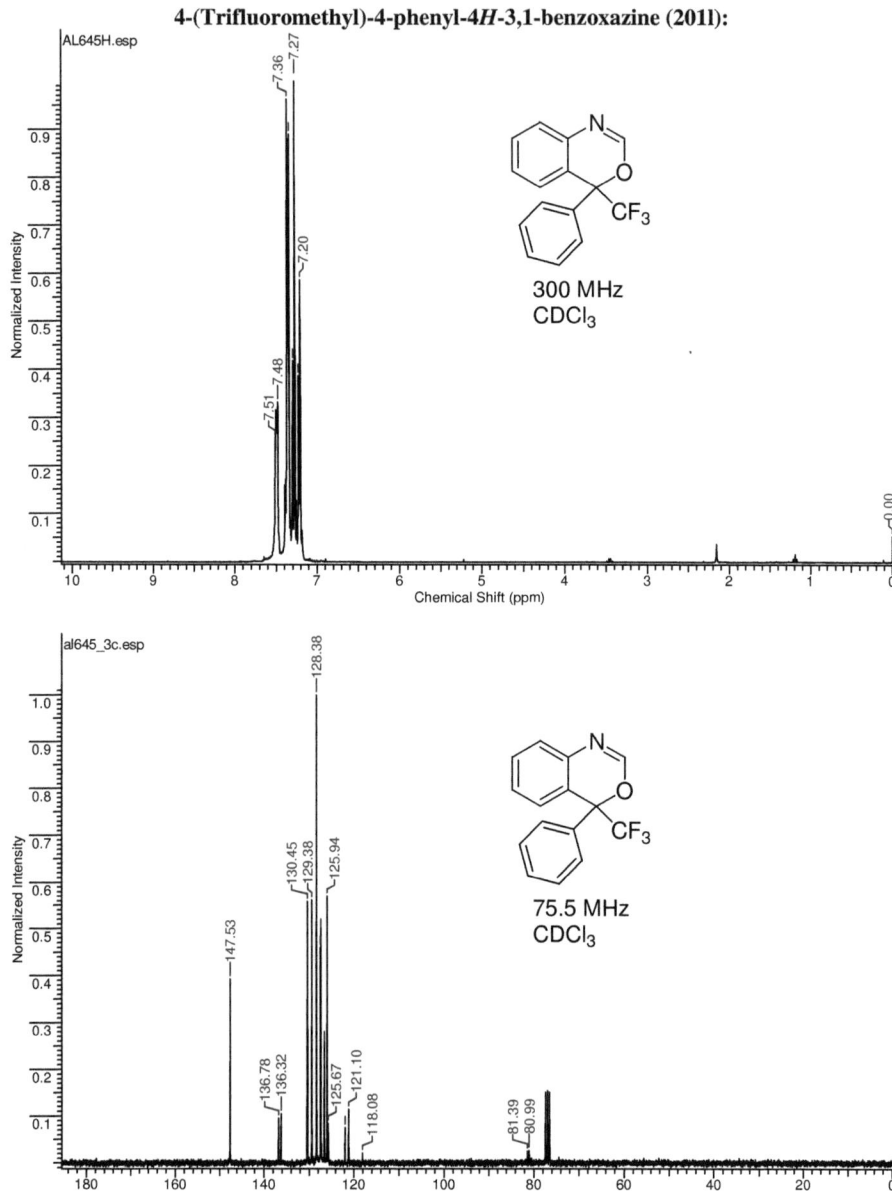

4-(Trifluoromethyl)-2-morpholino-4-phenyl-4*H*-3,1-benzoxazine (207)

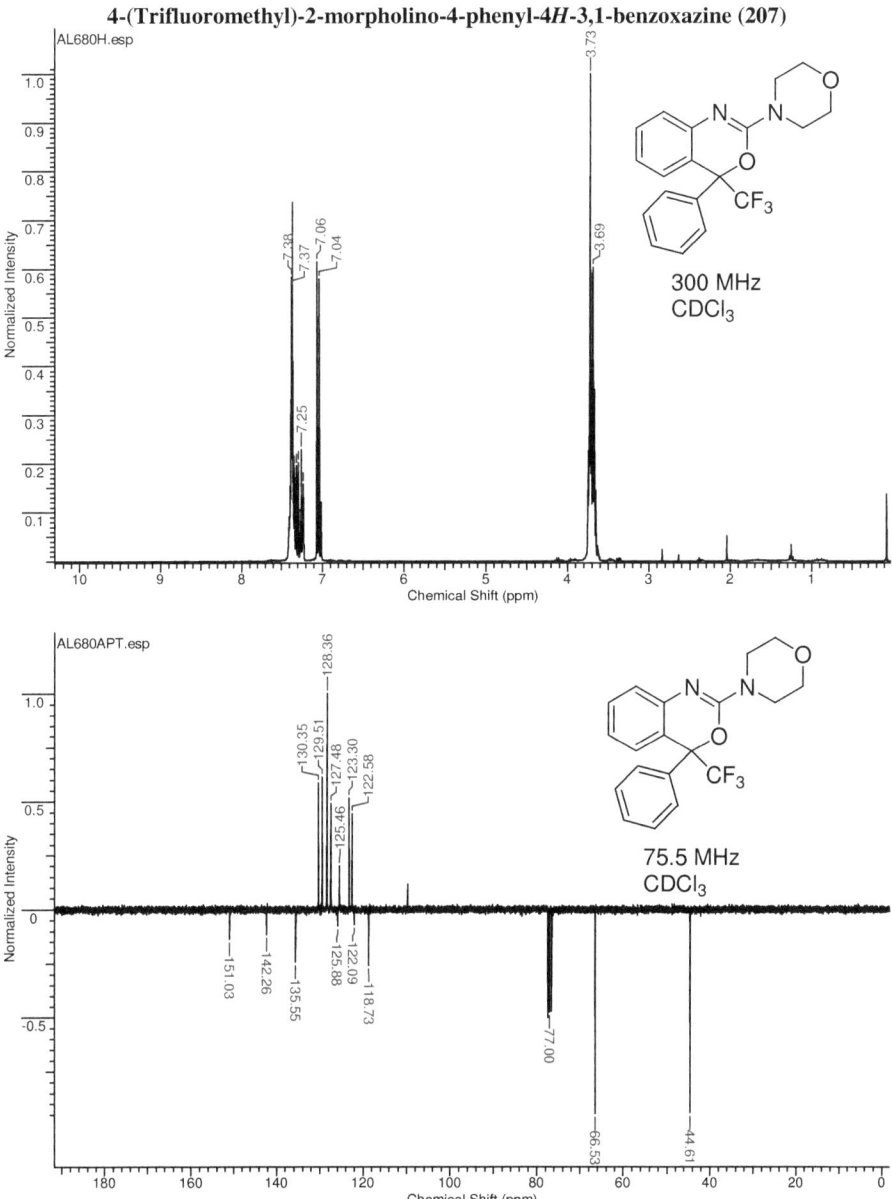

3-(Trifluoromethyl)-3-methylisobenzofuran-1(3H)-imine (210o)

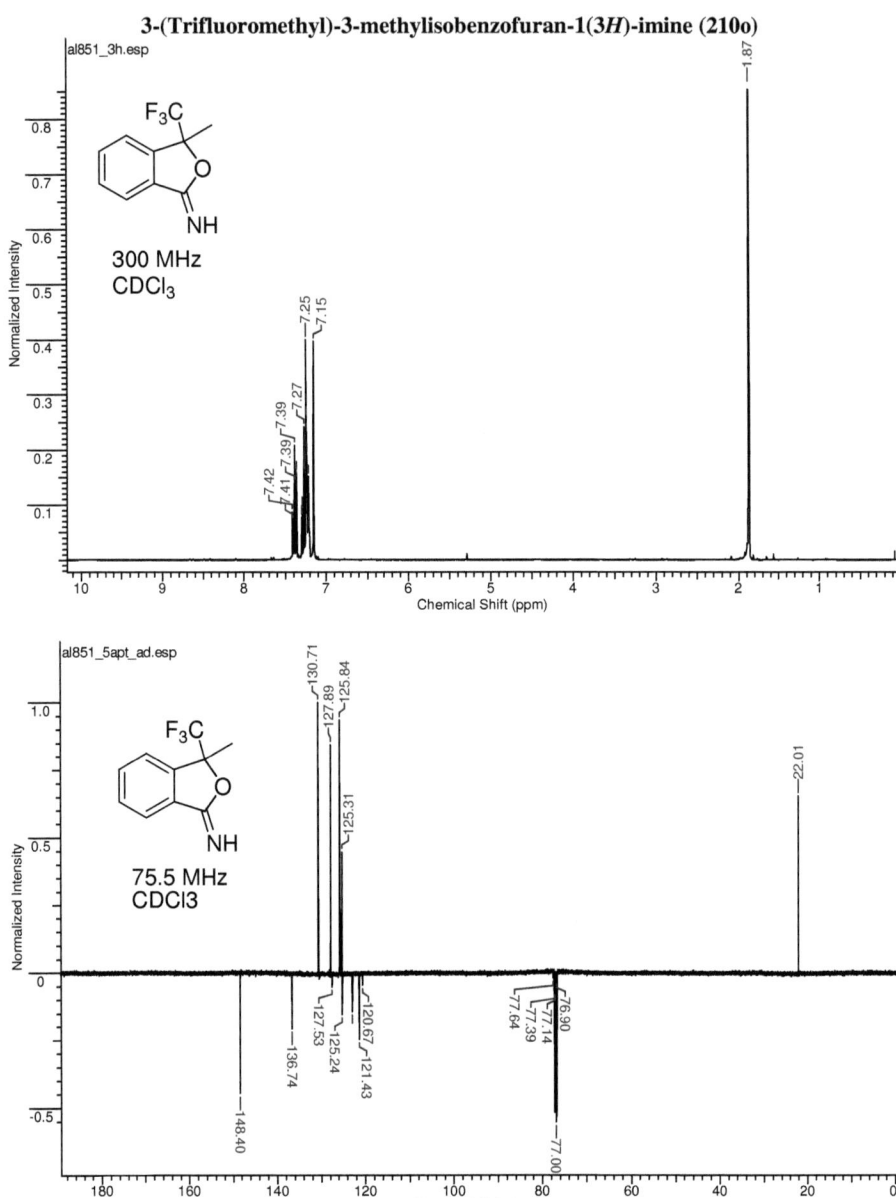

4-*tert*-Butyl-4*H*-3,1-benzoxazine (201g)

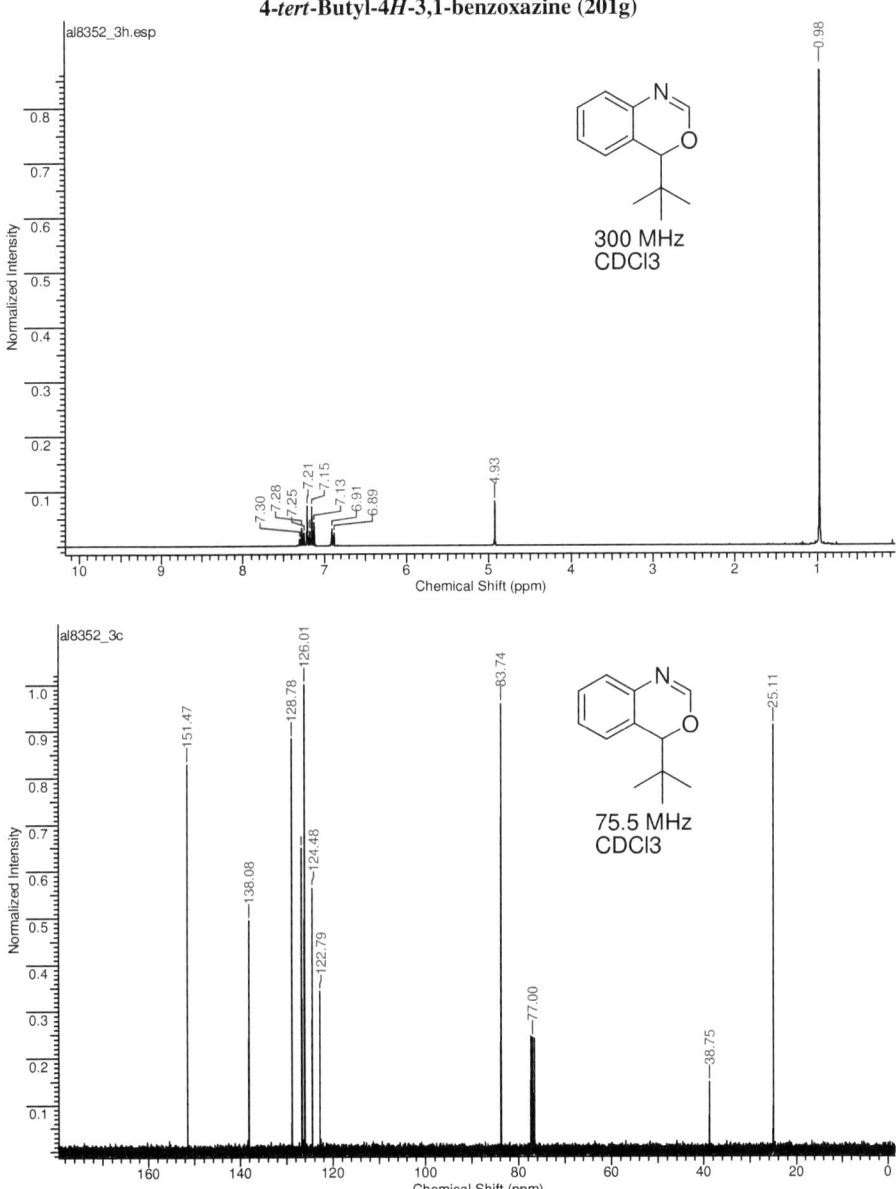

6,6-Diphenyl-4H-thieno[3,2-b]pyrrol-5(6H)-one (217k)

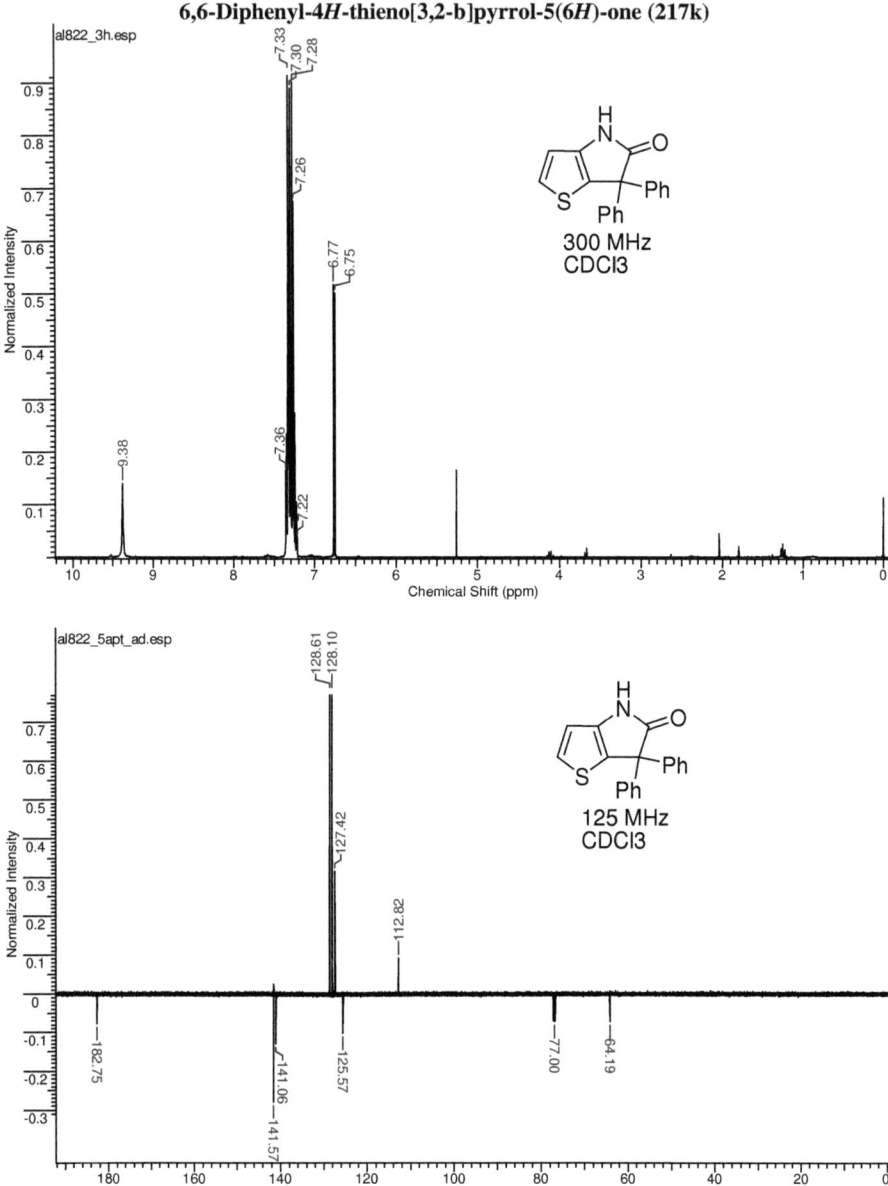

2-(Aziridin-1-yl)-4*H*-3,1-benzoxazin-4-one (199-azirid)

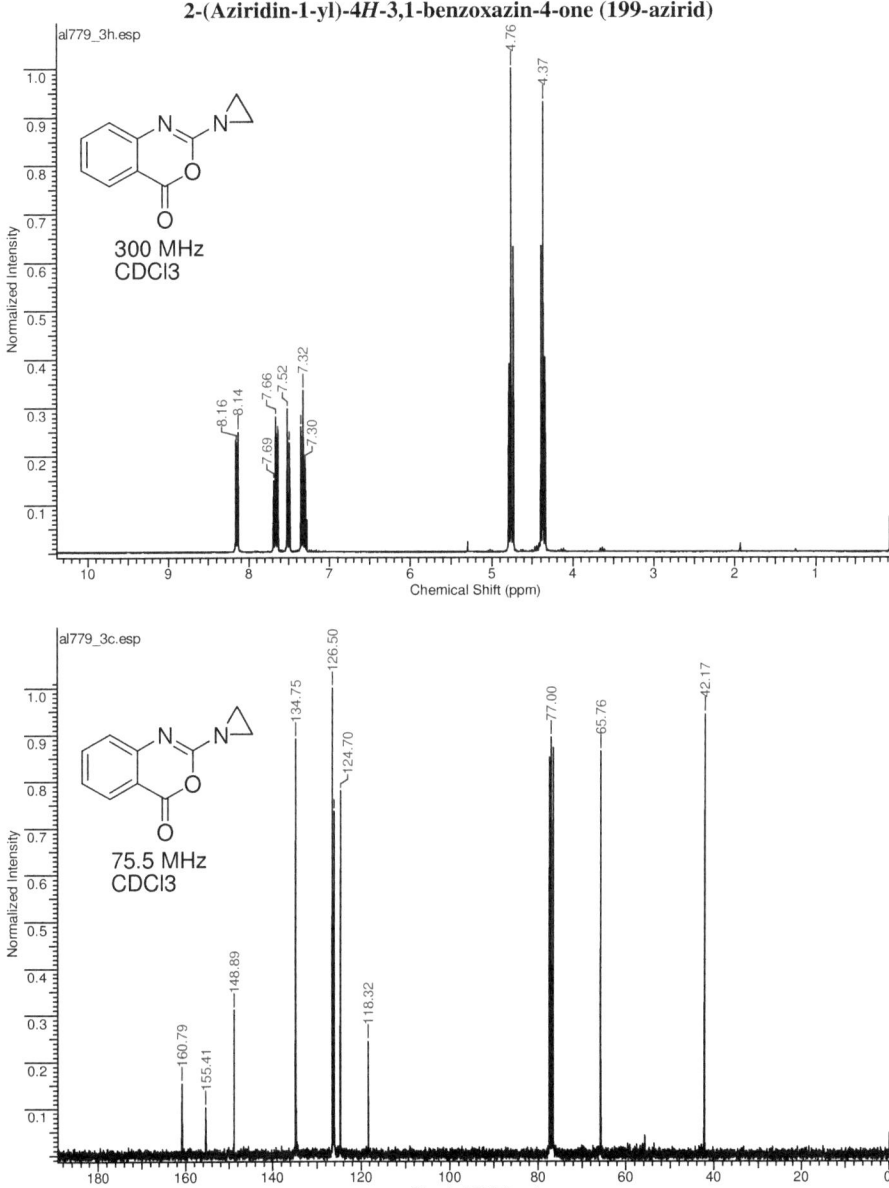

1-*n*-Propyl-1*H*-benzo[d]imidazole (232b)

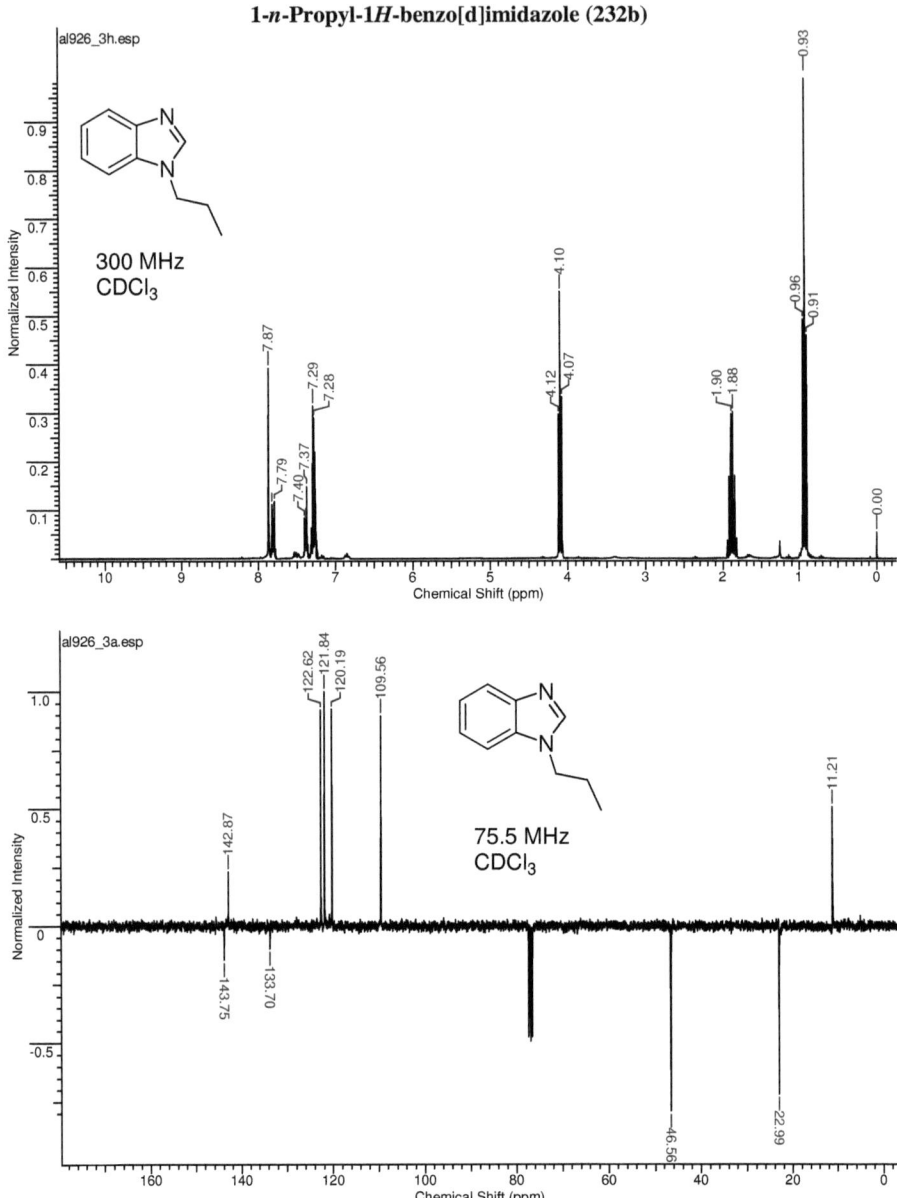

1-(2-(1*H*-Benzo[d]imidazol-1-yl)ethyl)-1*H*-benzo[d]imidazole (232e)

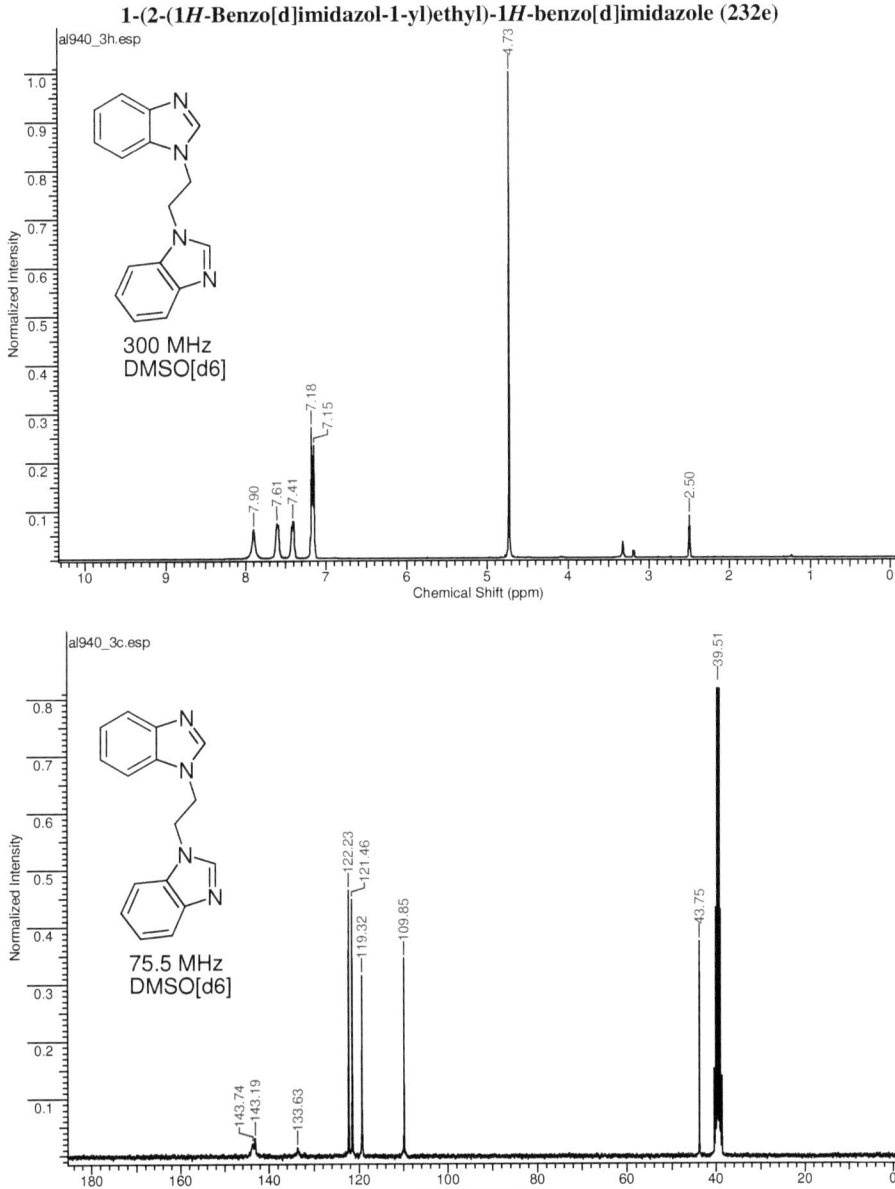

1-(2-(1-Methyl-1H-indol-3-yl)ethyl)-1H-benzo[d]imidazole (232d)

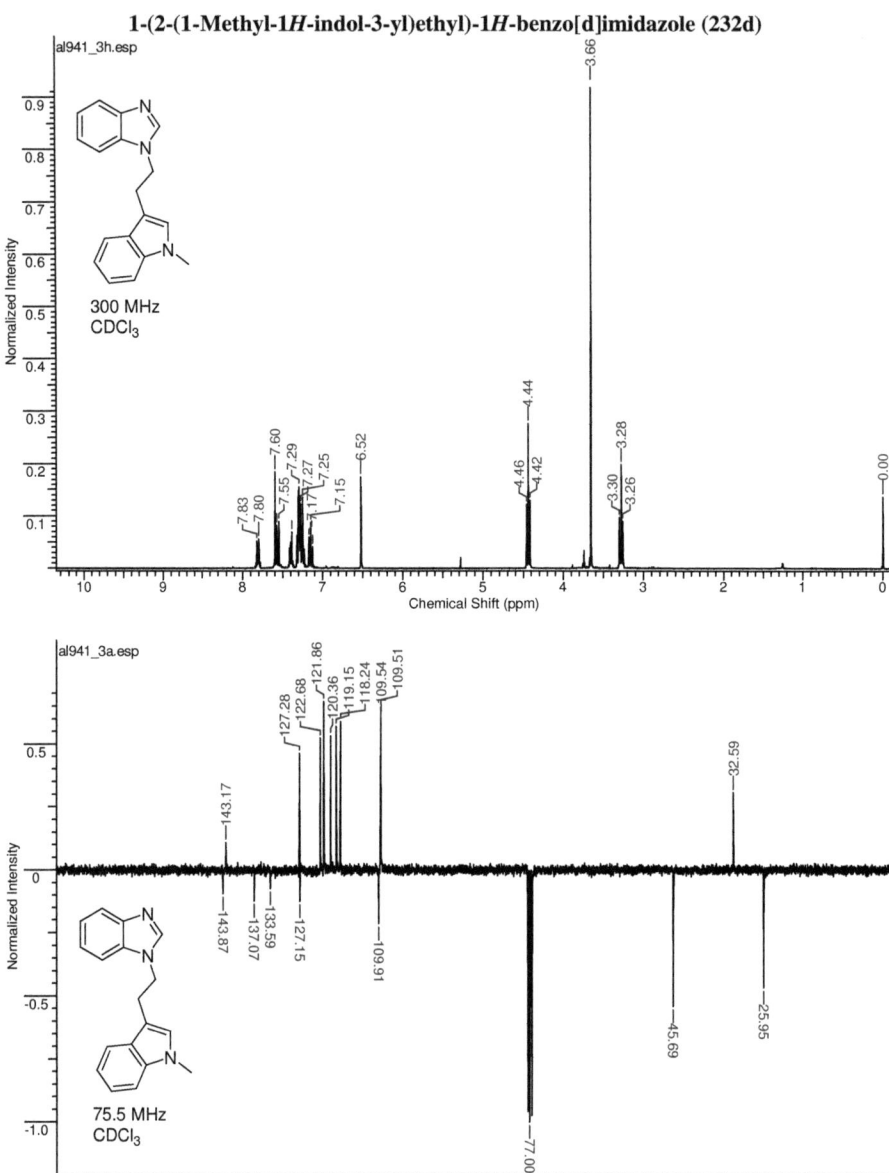

3-Benzyl-3*H*-thieno[2,3-d]imidazole (235a)

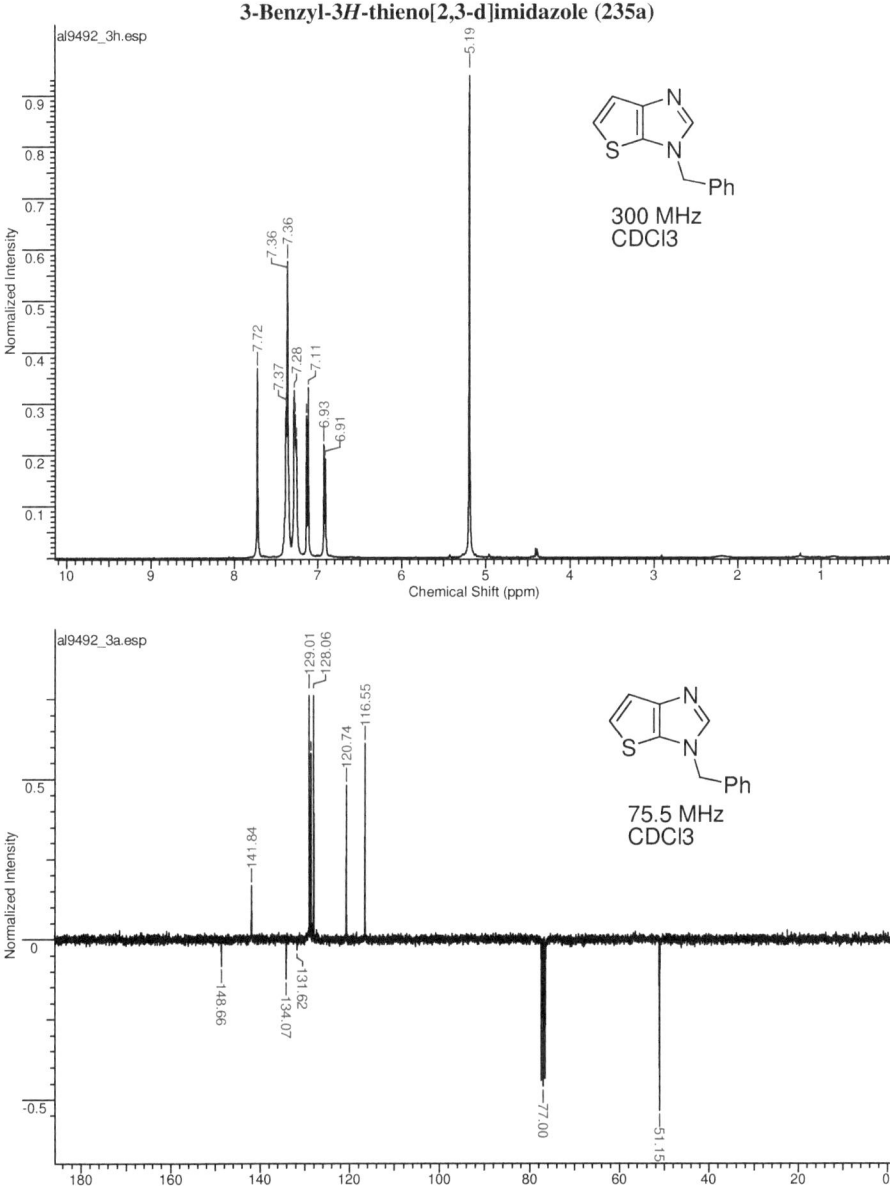

Acknowledgements

First of all, I thank my superviser and teacher Prof. Dr. A. de Meijere for giving me an opportunity to work on this thesis in Göttingen, for permanent support during the work and for his priceless help in preparation and correction of our common published works. The financial support of my participation in European-Asean Symposium in Obernai, France by Prof. de Meijere is also gratefully acknowledged.

I express my cordial gratitude and dedicate this work to my wife Tonja and daughter Masha, who have always been supported me on my way.

I am indebted to Degussa (Evonik)-Stiftung for a graduate student fellowship (2008-2009) and personally to Prof. Dr. W. Leuchtenberger, Dr. F. Sholl, Mrs. S. Peitzmann and Mrs. E. Sicht for perfect organisation of their work and help.

I thank Prof. Dr. de Meijere and Dr. M. Es-Sayed (Bayer Crop Science AG) for financial support at the beginning of my work in Göttingen (2006-2007).

I am also gratefull to Dr. M. Es-Sayed for his usefull advises and our fruitful collaboration.

I would like to say the words of gratitude to all people, with whom I have ever worked together in the de Meijere's group and who have shared with me their great experience and knowledge: Vadim Korotkov, Vitaliy Raev, Shamil Nizamov, Sergey M. Korneev, Dmitry Zabolotnev, Daniel Frank, Heiko Shill, Irina Martynova, Nikolai Ulin, Anna Osipova, Vladimir N. Belov, Sergey I. Kozhushkov, Viktor Bagutski, Andrey Savchenko and all short-time visitors.

For the unvaluable help with internet, software and all kinds of problems with computers I thank Heiko Shill, Daniel Frank and Stefan Beußhausen.

I am grateful to Alexey Nizovtzev for careful proofreading of this thesis.

I thank the former members of the group, Oleg Larionov and Tine Graef, with whom we had some fruitful collaborations concerning pyrrol syntheses (Oleg) and syntheses of Belactosine C analogues (Tine). I am also gratefull to Oleg and Vadim for discussions, motivation and co-working on the publications.

I am very thankful to Prof. Dr. U. Diederichsen for co-refering this thesis.

I appreciate the lectures and excersises held by Prof. Dr. F. Meyer, Prof. Dr. L. Ackermann, Prof. Dr. M. Buback, Dr. H.-P. Vogele, Prof. Dr. H. Laatsch that I attended and thank all these people for making their subjects clear and interesting.

I thank Prof. Dr. F. Meyer and Prof. Dr. M. Buback for being my examenators on subsidiary subjects "Catalysis" and "Technical Chemistry", respectively.

I am gratefull to Prof. Dr. H. Laatsch, Prof. Dr. S. Tsogoeva, Prof. Dr. M. Suhm, Prof. Dr. U. Klingebiel for acceptance of my diploma thesis from the M. V. Lomonosov Moscow State University and admitting so this study.

I would like to express my sincere gratitude to Mrs. G. Keil-Knepel for a lot of organisational problems she helped me to solve.

Some words of gratitude, which I must say about the staff of Institut für Organishe und Biomolekulare Chemie: I am indebted to Dipl.-Chem. R. Machinek and all his team as well as Dr. H. Frauendorf and Mrs. G. Udvarnoki for their fast and irreproachable work measuring NMR-spectra and mass-spectra, respectively. I thank Mr. F. Hambloch for the measurements of elemental analyses, Dipl.-Chem. O. Senge for the assistance by work with HPLC, Mrs. E. Pfeil for measurements of optical rotations and some of IR spectra. I am also grateful to Mr. R. Schrommek, Mr. H. Tucholla and Daniel Frank for the good organisation of the work and convenient supply of chemicals.

Special thanks to some of my friends, who helped me feel myself more comfortable in Germany, to learn German and not forget Russian: Katarina and Robert, Michael and Gelja, Vika and Vaciliy, Artyom and others.

Thank you all very much!!!

VDM Verlagsservicegesellschaft mbH

Die VDM Verlagsservicegesellschaft sucht für wissenschaftliche Verlage abgeschlossene und herausragende

Dissertationen, Habilitationen, Diplomarbeiten, Master Theses, Magisterarbeiten usw.

für die kostenlose Publikation als Fachbuch.

Sie verfügen über eine Arbeit, die hohen inhaltlichen und formalen Ansprüchen genügt, und haben Interesse an einer honorarvergüteten Publikation?

Dann senden Sie bitte erste Informationen über sich und Ihre Arbeit per Email an *info@vdm-vsg.de*.

Sie erhalten kurzfristig unser Feedback!

VDM Verlagsservicegesellschaft mbH
Dudweiler Landstr. 99
D - 66123 Saarbrücken
Telefon +49 681 3720 174
Fax +49 681 3720 1749
www.vdm-vsg.de

Die VDM Verlagsservicegesellschaft mbH vertritt

Printed by Books on Demand GmbH, Norderstedt / Germany